2007
Plasma Formulary

Prepared By

J.D. Huba

Beam Physics Branch
Plasma Physics Division

NRL

Wexford Press
2008

CONTENTS

NUMERICAL AND ALGEBRAIC

Gain in decibels of P_2 relative to P_1

$$G = 10 \log_{10}(P_2/P_1).$$

To within two percent

$$(2\pi)^{1/2} \approx 2.5; \quad \pi^2 \approx 10; \quad e^3 \approx 20; \quad 2^{10} \approx 10^3.$$

Euler-Mascheroni constant[1] $\gamma = 0.57722$

Gamma Function $\Gamma(x + 1) = x\Gamma(x)$:

$\Gamma(1/6) = 5.5663$	$\Gamma(3/5) = 1.4892$
$\Gamma(1/5) = 4.5908$	$\Gamma(2/3) = 1.3541$
$\Gamma(1/4) = 3.6256$	$\Gamma(3/4) = 1.2254$
$\Gamma(1/3) = 2.6789$	$\Gamma(4/5) = 1.1642$
$\Gamma(2/5) = 2.2182$	$\Gamma(5/6) = 1.1288$
$\Gamma(1/2) = 1.7725 = \sqrt{\pi}$	$\Gamma(1) = 1.0$

Binomial Theorem (good for $|x| < 1$ or $\alpha =$ positive integer):

$$(1 + x)^\alpha = \sum_{k=0}^{\infty} \binom{\alpha}{k} x^k \equiv 1 + \alpha x + \frac{\alpha(\alpha - 1)}{2!} x^2 + \frac{\alpha(\alpha - 1)(\alpha - 2)}{3!} x^3 + \dots.$$

Rothe-Hagen identity[2] (good for all complex x, y, z except when singular):

$$\sum_{k=0}^{n} \frac{x}{x + kz} \binom{x + kz}{k} \frac{y}{y + (n - k)z} \binom{y + (n - k)z}{n - k}$$

$$= \frac{x + y}{x + y + nz} \binom{x + y + nz}{n}.$$

Newberger's summation formula[3] [good for μ nonintegral, $\mathrm{Re}(\alpha + \beta) > -1$]:

$$\sum_{n=-\infty}^{\infty} \frac{(-1)^n J_{\alpha-\gamma n}(z) J_{\beta+\gamma n}(z)}{n + \mu} = \frac{\pi}{\sin \mu\pi} J_{\alpha+\gamma\mu}(z) J_{\beta-\gamma\mu}(z).$$

Notation: f, g, are scalars; $\mathbf{A}$, $\mathbf{B}$, etc., are vectors; T is a tensor; I is the unit dyad.

(1) $\mathbf{A} \cdot \mathbf{B} \times \mathbf{C} = \mathbf{A} \times \mathbf{B} \cdot \mathbf{C} = \mathbf{B} \cdot \mathbf{C} \times \mathbf{A} = \mathbf{B} \times \mathbf{C} \cdot \mathbf{A} = \mathbf{C} \cdot \mathbf{A} \times \mathbf{B} = \mathbf{C} \times \mathbf{A} \cdot \mathbf{B}$

(2) $\mathbf{A} \times (\mathbf{B} \times \mathbf{C}) = (\mathbf{C} \times \mathbf{B}) \times \mathbf{A} = (\mathbf{A} \cdot \mathbf{C})\mathbf{B} - (\mathbf{A} \cdot \mathbf{B})\mathbf{C}$

(3) $\mathbf{A} \times (\mathbf{B} \times \mathbf{C}) + \mathbf{B} \times (\mathbf{C} \times \mathbf{A}) + \mathbf{C} \times (\mathbf{A} \times \mathbf{B}) = 0$

(4) $(\mathbf{A} \times \mathbf{B}) \cdot (\mathbf{C} \times \mathbf{D}) = (\mathbf{A} \cdot \mathbf{C})(\mathbf{B} \cdot \mathbf{D}) - (\mathbf{A} \cdot \mathbf{D})(\mathbf{B} \cdot \mathbf{C})$

(5) $(\mathbf{A} \times \mathbf{B}) \times (\mathbf{C} \times \mathbf{D}) = (\mathbf{A} \times \mathbf{B} \cdot \mathbf{D})\mathbf{C} - (\mathbf{A} \times \mathbf{B} \cdot \mathbf{C})\mathbf{D}$

(6) $\nabla(fg) = \nabla(gf) = f\nabla g + g\nabla f$

(7) $\nabla \cdot (f\mathbf{A}) = f\nabla \cdot \mathbf{A} + \mathbf{A} \cdot \nabla f$

(8) $\nabla \times (f\mathbf{A}) = f\nabla \times \mathbf{A} + \nabla f \times \mathbf{A}$

(9) $\nabla \cdot (\mathbf{A} \times \mathbf{B}) = \mathbf{B} \cdot \nabla \times \mathbf{A} - \mathbf{A} \cdot \nabla \times \mathbf{B}$

(10) $\nabla \times (\mathbf{A} \times \mathbf{B}) = \mathbf{A}(\nabla \cdot \mathbf{B}) - \mathbf{B}(\nabla \cdot \mathbf{A}) + (\mathbf{B} \cdot \nabla)\mathbf{A} - (\mathbf{A} \cdot \nabla)\mathbf{B}$

(11) $\mathbf{A} \times (\nabla \times \mathbf{B}) = (\nabla\mathbf{B}) \cdot \mathbf{A} - (\mathbf{A} \cdot \nabla)\mathbf{B}$

(12) $\nabla(\mathbf{A} \cdot \mathbf{B}) = \mathbf{A} \times (\nabla \times \mathbf{B}) + \mathbf{B} \times (\nabla \times \mathbf{A}) + (\mathbf{A} \cdot \nabla)\mathbf{B} + (\mathbf{B} \cdot \nabla)\mathbf{A}$

(13) $\nabla^2 f = \nabla \cdot \nabla f$

(14) $\nabla^2 \mathbf{A} = \nabla(\nabla \cdot \mathbf{A}) - \nabla \times \nabla \times \mathbf{A}$

(15) $\nabla \times \nabla f = 0$

(16) $\nabla \cdot \nabla \times \mathbf{A} = 0$

If $\mathbf{e}_1$, $\mathbf{e}_2$, $\mathbf{e}_3$ are orthonormal unit vectors, a second-order tensor T can be written in the dyadic form

(17) $\mathsf{T} = \sum_{i,j} T_{ij}\mathbf{e}_i\mathbf{e}_j$

In cartesian coordinates the divergence of a tensor is a vector with components

(18) $(\nabla \cdot \mathsf{T})_i = \sum_j (\partial T_{ji}/\partial x_j)$

[This definition is required for consistency with Eq. (29)]. In general

(19) $\nabla \cdot (\mathbf{A}\mathbf{B}) = (\nabla \cdot \mathbf{A})\mathbf{B} + (\mathbf{A} \cdot \nabla)\mathbf{B}$

(20) $\nabla \cdot (f\mathsf{T}) = \nabla f \cdot \mathsf{T} + f\nabla \cdot \mathsf{T}$

4

Let $\mathbf{r} = \mathbf{i}x + \mathbf{j}y + \mathbf{k}z$ be the radius vector of magnitude r, from the origin to the point x, y, z. Then

(21) $\nabla \cdot \mathbf{r} = 3$

(22) $\nabla \times \mathbf{r} = 0$

(23) $\nabla r = \mathbf{r}/r$

(24) $\nabla(1/r) = -\mathbf{r}/r^3$

(25) $\nabla \cdot (\mathbf{r}/r^3) = 4\pi\delta(\mathbf{r})$

(26) $\nabla\mathbf{r} = \mathit{I}$

If V is a volume enclosed by a surface S and $d\mathbf{S} = \mathbf{n}dS$, where $\mathbf{n}$ is the unit normal outward from V,

(27) $\displaystyle\int_V dV\,\nabla f = \int_S d\mathbf{S}\,f$

(28) $\displaystyle\int_V dV\,\nabla \cdot \mathbf{A} = \int_S d\mathbf{S} \cdot \mathbf{A}$

(29) $\displaystyle\int_V dV\,\nabla \cdot \mathit{T} = \int_S d\mathbf{S} \cdot \mathit{T}$

(30) $\displaystyle\int_V dV\,\nabla \times \mathbf{A} = \int_S d\mathbf{S} \times \mathbf{A}$

(31) $\displaystyle\int_V dV\,(f\nabla^2 g - g\nabla^2 f) = \int_S d\mathbf{S} \cdot (f\nabla g - g\nabla f)$

(32) $\displaystyle\int_V dV\,(\mathbf{A} \cdot \nabla \times \nabla \times \mathbf{B} - \mathbf{B} \cdot \nabla \times \nabla \times \mathbf{A})$

$$= \int_S d\mathbf{S} \cdot (\mathbf{B} \times \nabla \times \mathbf{A} - \mathbf{A} \times \nabla \times \mathbf{B})$$

If S is an open surface bounded by the contour C, of which the line element is $d\mathbf{l}$,

(33) $\displaystyle\int_S d\mathbf{S} \times \nabla f = \oint_C d\mathbf{l}\,f$

$$(34) \quad \int_S d\mathbf{S} \cdot \nabla \times \mathbf{A} = \oint_C d\mathbf{l} \cdot \mathbf{A}$$

$$(35) \quad \int_S (d\mathbf{S} \times \nabla) \times \mathbf{A} = \oint_C d\mathbf{l} \times \mathbf{A}$$

$$(36) \quad \int_S d\mathbf{S} \cdot (\nabla f \times \nabla g) = \oint_C f \, dg = - \oint_C g \, df$$

DIFFERENTIAL OPERATORS IN CURVILINEAR COORDINATES[5]

Cylindrical Coordinates

Divergence

$$\nabla \cdot \mathbf{A} = \frac{1}{r}\frac{\partial}{\partial r}(rA_r) + \frac{1}{r}\frac{\partial A_\phi}{\partial \phi} + \frac{\partial A_z}{\partial z}$$

Gradient

$$(\nabla f)_r = \frac{\partial f}{\partial r}; \quad (\nabla f)_\phi = \frac{1}{r}\frac{\partial f}{\partial \phi}; \quad (\nabla f)_z = \frac{\partial f}{\partial z}$$

Curl

$$(\nabla \times \mathbf{A})_r = \frac{1}{r}\frac{\partial A_z}{\partial \phi} - \frac{\partial A_\phi}{\partial z}$$

$$(\nabla \times \mathbf{A})_\phi = \frac{\partial A_r}{\partial z} - \frac{\partial A_z}{\partial r}$$

$$(\nabla \times \mathbf{A})_z = \frac{1}{r}\frac{\partial}{\partial r}(rA_\phi) - \frac{1}{r}\frac{\partial A_r}{\partial \phi}$$

Laplacian

$$\nabla^2 f = \frac{1}{r}\frac{\partial}{\partial r}\left(r\frac{\partial f}{\partial r}\right) + \frac{1}{r^2}\frac{\partial^2 f}{\partial \phi^2} + \frac{\partial^2 f}{\partial z^2}$$

6

Laplacian of a vector

$$(\nabla^2 \mathbf{A})_r = \nabla^2 A_r - \frac{2}{r^2}\frac{\partial A_\phi}{\partial \phi} - \frac{A_r}{r^2}$$

$$(\nabla^2 \mathbf{A})_\phi = \nabla^2 A_\phi + \frac{2}{r^2}\frac{\partial A_r}{\partial \phi} - \frac{A_\phi}{r^2}$$

$$(\nabla^2 \mathbf{A})_z = \nabla^2 A_z$$

Components of $(\mathbf{A} \cdot \nabla)\mathbf{B}$

$$(\mathbf{A} \cdot \nabla \mathbf{B})_r = A_r\frac{\partial B_r}{\partial r} + \frac{A_\phi}{r}\frac{\partial B_r}{\partial \phi} + A_z\frac{\partial B_r}{\partial z} - \frac{A_\phi B_\phi}{r}$$

$$(\mathbf{A} \cdot \nabla \mathbf{B})_\phi = A_r\frac{\partial B_\phi}{\partial r} + \frac{A_\phi}{r}\frac{\partial B_\phi}{\partial \phi} + A_z\frac{\partial B_\phi}{\partial z} + \frac{A_\phi B_r}{r}$$

$$(\mathbf{A} \cdot \nabla \mathbf{B})_z = A_r\frac{\partial B_z}{\partial r} + \frac{A_\phi}{r}\frac{\partial B_z}{\partial \phi} + A_z\frac{\partial B_z}{\partial z}$$

Divergence of a tensor

$$(\nabla \cdot T)_r = \frac{1}{r}\frac{\partial}{\partial r}(rT_{rr}) + \frac{1}{r}\frac{\partial T_{\phi r}}{\partial \phi} + \frac{\partial T_{zr}}{\partial z} - \frac{T_{\phi\phi}}{r}$$

$$(\nabla \cdot T)_\phi = \frac{1}{r}\frac{\partial}{\partial r}(rT_{r\phi}) + \frac{1}{r}\frac{\partial T_{\phi\phi}}{\partial \phi} + \frac{\partial T_{z\phi}}{\partial z} + \frac{T_{\phi r}}{r}$$

$$(\nabla \cdot T)_z = \frac{1}{r}\frac{\partial}{\partial r}(rT_{rz}) + \frac{1}{r}\frac{\partial T_{\phi z}}{\partial \phi} + \frac{\partial T_{zz}}{\partial z}$$

Spherical Coordinates

Divergence

$$\nabla \cdot \mathbf{A} = \frac{1}{r^2} \frac{\partial}{\partial r} (r^2 A_r) + \frac{1}{r \sin \theta} \frac{\partial}{\partial \theta} (\sin \theta A_\theta) + \frac{1}{r \sin \theta} \frac{\partial A_\phi}{\partial \phi}$$

Gradient

$$(\nabla f)_r = \frac{\partial f}{\partial r}; \quad (\nabla f)_\theta = \frac{1}{r} \frac{\partial f}{\partial \theta}; \quad (\nabla f)_\phi = \frac{1}{r \sin \theta} \frac{\partial f}{\partial \phi}$$

Curl

$$(\nabla \times \mathbf{A})_r = \frac{1}{r \sin \theta} \frac{\partial}{\partial \theta} (\sin \theta A_\phi) - \frac{1}{r \sin \theta} \frac{\partial A_\theta}{\partial \phi}$$

$$(\nabla \times \mathbf{A})_\theta = \frac{1}{r \sin \theta} \frac{\partial A_r}{\partial \phi} - \frac{1}{r} \frac{\partial}{\partial r} (r A_\phi)$$

$$(\nabla \times \mathbf{A})_\phi = \frac{1}{r} \frac{\partial}{\partial r} (r A_\theta) - \frac{1}{r} \frac{\partial A_r}{\partial \theta}$$

Laplacian

$$\nabla^2 f = \frac{1}{r^2} \frac{\partial}{\partial r} \left(r^2 \frac{\partial f}{\partial r} \right) + \frac{1}{r^2 \sin \theta} \frac{\partial}{\partial \theta} \left(\sin \theta \frac{\partial f}{\partial \theta} \right) + \frac{1}{r^2 \sin^2 \theta} \frac{\partial^2 f}{\partial \phi^2}$$

Laplacian of a vector

$$(\nabla^2 \mathbf{A})_r = \nabla^2 A_r - \frac{2 A_r}{r^2} - \frac{2}{r^2} \frac{\partial A_\theta}{\partial \theta} - \frac{2 \cot \theta A_\theta}{r^2} - \frac{2}{r^2 \sin \theta} \frac{\partial A_\phi}{\partial \phi}$$

$$(\nabla^2 \mathbf{A})_\theta = \nabla^2 A_\theta + \frac{2}{r^2} \frac{\partial A_r}{\partial \theta} - \frac{A_\theta}{r^2 \sin^2 \theta} - \frac{2 \cos \theta}{r^2 \sin^2 \theta} \frac{\partial A_\phi}{\partial \phi}$$

$$(\nabla^2 \mathbf{A})_\phi = \nabla^2 A_\phi - \frac{A_\phi}{r^2 \sin^2 \theta} + \frac{2}{r^2 \sin \theta} \frac{\partial A_r}{\partial \phi} + \frac{2 \cos \theta}{r^2 \sin^2 \theta} \frac{\partial A_\theta}{\partial \phi}$$

Components of $(\mathbf{A} \cdot \nabla)\mathbf{B}$

$$(\mathbf{A} \cdot \nabla \mathbf{B})_r = A_r \frac{\partial B_r}{\partial r} + \frac{A_\theta}{r} \frac{\partial B_r}{\partial \theta} + \frac{A_\phi}{r \sin \theta} \frac{\partial B_r}{\partial \phi} - \frac{A_\theta B_\theta + A_\phi B_\phi}{r}$$

$$(\mathbf{A} \cdot \nabla \mathbf{B})_\theta = A_r \frac{\partial B_\theta}{\partial r} + \frac{A_\theta}{r} \frac{\partial B_\theta}{\partial \theta} + \frac{A_\phi}{r \sin \theta} \frac{\partial B_\theta}{\partial \phi} + \frac{A_\theta B_r}{r} - \frac{\cot \theta A_\phi B_\phi}{r}$$

$$(\mathbf{A} \cdot \nabla \mathbf{B})_\phi = A_r \frac{\partial B_\phi}{\partial r} + \frac{A_\theta}{r} \frac{\partial B_\phi}{\partial \theta} + \frac{A_\phi}{r \sin \theta} \frac{\partial B_\phi}{\partial \phi} + \frac{A_\phi B_r}{r} + \frac{\cot \theta A_\phi B_\theta}{r}$$

Divergence of a tensor

$$(\nabla \cdot T)_r = \frac{1}{r^2} \frac{\partial}{\partial r}(r^2 T_{rr}) + \frac{1}{r \sin \theta} \frac{\partial}{\partial \theta}(\sin \theta T_{\theta r})$$

$$+ \frac{1}{r \sin \theta} \frac{\partial T_{\phi r}}{\partial \phi} - \frac{T_{\theta\theta} + T_{\phi\phi}}{r}$$

$$(\nabla \cdot T)_\theta = \frac{1}{r^2} \frac{\partial}{\partial r}(r^2 T_{r\theta}) + \frac{1}{r \sin \theta} \frac{\partial}{\partial \theta}(\sin \theta T_{\theta\theta})$$

$$+ \frac{1}{r \sin \theta} \frac{\partial T_{\phi\theta}}{\partial \phi} + \frac{T_{\theta r}}{r} - \frac{\cot \theta T_{\phi\phi}}{r}$$

$$(\nabla \cdot T)_\phi = \frac{1}{r^2} \frac{\partial}{\partial r}(r^2 T_{r\phi}) + \frac{1}{r \sin \theta} \frac{\partial}{\partial \theta}(\sin \theta T_{\theta\phi})$$

$$+ \frac{1}{r \sin \theta} \frac{\partial T_{\phi\phi}}{\partial \phi} + \frac{T_{\phi r}}{r} + \frac{\cot \theta T_{\phi\theta}}{r}$$

DIMENSIONS AND UNITS

To get the value of a quantity in Gaussian units, multiply the value expressed in SI units by the conversion factor. Multiples of 3 in the conversion factors result from approximating the speed of light $c = 2.9979 \times 10^{10}$ cm/sec $\approx 3 \times 10^{10}$ cm/sec.

Physical Quantity	Symbol	Dimensions		SI Units	Conversion Factor	Gaussian Units
		SI	Gaussian			
Capacitance	C	$\dfrac{t^2 q^2}{ml^2}$	l	farad	9×10^{11}	cm
Charge	q	q	$\dfrac{m^{1/2} l^{3/2}}{t}$	coulomb	3×10^9	statcoulomb
Charge density	ρ	$\dfrac{q}{l^3}$	$\dfrac{m^{1/2}}{l^{3/2} t}$	coulomb /m^3	3×10^3	statcoulomb /cm^3
Conductance		$\dfrac{t q^2}{ml^2}$	$\dfrac{l}{t}$	siemens	9×10^{11}	cm/sec
Conductivity	σ	$\dfrac{t q^2}{ml^3}$	$\dfrac{1}{t}$	siemens /m	9×10^9	sec^{-1}
Current	I, i	$\dfrac{q}{t}$	$\dfrac{m^{1/2} l^{3/2}}{t^2}$	ampere	3×10^9	statampere
Current density	$\mathbf{J}, \mathbf{j}$	$\dfrac{q}{l^2 t}$	$\dfrac{m^{1/2}}{l^{1/2} t^2}$	ampere /m^2	3×10^5	statampere /cm^2
Density	ρ	$\dfrac{m}{l^3}$	$\dfrac{m}{l^3}$	kg/m^3	10^{-3}	g/cm^3
Displacement	$\mathbf{D}$	$\dfrac{q}{l^2}$	$\dfrac{m^{1/2}}{l^{1/2} t}$	coulomb /m^2	$12\pi \times 10^5$	statcoulomb /cm^2
Electric field	$\mathbf{E}$	$\dfrac{ml}{t^2 q}$	$\dfrac{m^{1/2}}{l^{1/2} t}$	volt/m	$\dfrac{1}{3} \times 10^{-4}$	statvolt/cm
Electromotance	$\mathcal{E}$, Emf	$\dfrac{ml^2}{t^2 q}$	$\dfrac{m^{1/2} l^{1/2}}{t}$	volt	$\dfrac{1}{3} \times 10^{-2}$	statvolt
Energy	U, W	$\dfrac{ml^2}{t^2}$	$\dfrac{ml^2}{t^2}$	joule	10^7	erg
Energy density	w, ϵ	$\dfrac{m}{lt^2}$	$\dfrac{m}{lt^2}$	joule/m^3	10	erg/cm^3

10

Physical Quantity	Symbol	Dimensions SI	Dimensions Gaussian	SI Units	Conversion Factor	Gaussian Units
Force	**F**	$\dfrac{ml}{t^2}$	$\dfrac{ml}{t^2}$	newton	10^5	dyne
Frequency	f, ν	$\dfrac{1}{t}$	$\dfrac{1}{t}$	hertz	1	hertz
Impedance	Z	$\dfrac{ml^2}{tq^2}$	$\dfrac{t}{l}$	ohm	$\dfrac{1}{9} \times 10^{-11}$	sec/cm
Inductance	L	$\dfrac{ml^2}{q^2}$	$\dfrac{t^2}{l}$	henry	$\dfrac{1}{9} \times 10^{-11}$	sec^2/cm
Length	l	l	l	meter (m)	10^2	centimeter (cm)
Magnetic intensity	**H**	$\dfrac{q}{lt}$	$\dfrac{m^{1/2}}{l^{1/2}t}$	ampere–turn/m	$4\pi \times 10^{-3}$	oersted
Magnetic flux	Φ	$\dfrac{ml^2}{tq}$	$\dfrac{m^{1/2}l^{3/2}}{t}$	weber	10^8	maxwell
Magnetic induction	**B**	$\dfrac{m}{tq}$	$\dfrac{m^{1/2}}{l^{1/2}t}$	tesla	10^4	gauss
Magnetic moment	m, μ	$\dfrac{l^2 q}{t}$	$\dfrac{m^{1/2}l^{5/2}}{t}$	ampere–m^2	10^3	oersted–cm^3
Magnetization	**M**	$\dfrac{q}{lt}$	$\dfrac{m^{1/2}}{l^{1/2}t}$	ampere–turn/m	$4\pi \times 10^{-3}$	oersted
Magneto-motance	$\mathcal{M}$, Mmf	$\dfrac{q}{t}$	$\dfrac{m^{1/2}l^{1/2}}{t^2}$	ampere–turn	$\dfrac{4\pi}{10}$	gilbert
Mass	m, M	m	m	kilogram (kg)	10^3	gram (g)
Momentum	**p, P**	$\dfrac{ml}{t}$	$\dfrac{ml}{t}$	kg–m/s	10^5	g–cm/sec
Momentum density		$\dfrac{m}{l^2 t}$	$\dfrac{m}{l^2 t}$	kg/m^2–s	10^{-1}	g/cm^2–sec
Permeability	μ	$\dfrac{ml}{q^2}$	1	henry/m	$\dfrac{1}{4\pi} \times 10^7$	—

Physical Quantity	Symbol	Dimensions		SI Units	Conversion Factor	Gaussian Units
		SI	Gaussian			
Permittivity	ϵ	$\dfrac{t^2 q^2}{ml^3}$	1	farad/m	$36\pi \times 10^9$	—
Polarization	**P**	$\dfrac{q}{l^2}$	$\dfrac{m^{1/2}}{l^{1/2}t}$	coulomb/m^2	3×10^5	statcoulomb /cm^2
Potential	V, ϕ	$\dfrac{ml^2}{t^2 q}$	$\dfrac{m^{1/2}l^{1/2}}{t}$	volt	$\dfrac{1}{3} \times 10^{-2}$	statvolt
Power	P	$\dfrac{ml^2}{t^3}$	$\dfrac{ml^2}{t^3}$	watt	10^7	erg/sec
Power density		$\dfrac{m}{lt^3}$	$\dfrac{m}{lt^3}$	watt/m^3	10	erg/cm^3–sec
Pressure	p, P	$\dfrac{m}{lt^2}$	$\dfrac{m}{lt^2}$	pascal	10	dyne/cm^2
Reluctance	$\mathcal{R}$	$\dfrac{q^2}{ml^2}$	$\dfrac{1}{l}$	ampere–turn /weber	$4\pi \times 10^{-9}$	cm^{-1}
Resistance	R	$\dfrac{ml^2}{tq^2}$	$\dfrac{t}{l}$	ohm	$\dfrac{1}{9} \times 10^{-11}$	sec/cm
Resistivity	η, ρ	$\dfrac{ml^3}{tq^2}$	t	ohm–m	$\dfrac{1}{9} \times 10^{-9}$	sec
Thermal conductivity	κ, k	$\dfrac{ml}{t^3}$	$\dfrac{ml}{t^3}$	watt/m–deg (K)	10^5	erg/cm–sec–deg (K)
Time	t	t	t	second (s)	1	second (sec)
Vector potential	**A**	$\dfrac{ml}{tq}$	$\dfrac{m^{1/2}l^{1/2}}{t}$	weber/m	10^6	gauss–cm
Velocity	**v**	$\dfrac{l}{t}$	$\dfrac{l}{t}$	m/s	10^2	cm/sec
Viscosity	η, μ	$\dfrac{m}{lt}$	$\dfrac{m}{lt}$	kg/m–s	10	poise
Vorticity	ζ	$\dfrac{1}{t}$	$\dfrac{1}{t}$	s^{-1}	1	sec^{-1}
Work	W	$\dfrac{ml^2}{t^2}$	$\dfrac{ml^2}{t^2}$	joule	10^7	erg

INTERNATIONAL SYSTEM (SI) NOMENCLATURE[6]

Physical Quantity	Name of Unit	Symbol for Unit	Physical Quantity	Name of Unit	Symbol for Unit
*length	meter	m	electric potential	volt	V
*mass	kilogram	kg	electric resistance	ohm	Ω
*time	second	s			
*current	ampere	A	electric conductance	siemens	S
*temperature	kelvin	K			
*amount of substance	mole	mol	electric capacitance	farad	F
			magnetic flux	weber	Wb
*luminous intensity	candela	cd	magnetic inductance	henry	H
†plane angle	radian	rad			
†solid angle	steradian	sr	magnetic intensity	tesla	T
frequency	hertz	Hz	luminous flux	lumen	lm
energy	joule	J	illuminance	lux	lx
force	newton	N	activity (of a radioactive source)	becquerel	Bq
pressure	pascal	Pa			
power	watt	W	absorbed dose (of ionizing radiation)	gray	Gy
electric charge	coulomb	C			

*SI base unit   †Supplementary unit

METRIC PREFIXES

Multiple	Prefix	Symbol	Multiple	Prefix	Symbol
10^{-1}	deci	d	10	deca	da
10^{-2}	centi	c	10^2	hecto	h
10^{-3}	milli	m	10^3	kilo	k
10^{-6}	micro	μ	10^6	mega	M
10^{-9}	nano	n	10^9	giga	G
10^{-12}	pico	p	10^{12}	tera	T
10^{-15}	femto	f	10^{15}	peta	P
10^{-18}	atto	a	10^{18}	exa	E

PHYSICAL CONSTANTS (SI)[7]

Physical Quantity	Symbol	Value	Units
Boltzmann constant	k	1.3807×10^{-23}	$\mathrm{J\,K^{-1}}$
Elementary charge	e	1.6022×10^{-19}	C
Electron mass	m_e	9.1094×10^{-31}	kg
Proton mass	m_p	1.6726×10^{-27}	kg
Gravitational constant	G	6.6726×10^{-11}	$\mathrm{m^3 s^{-2} kg^{-1}}$
Planck constant	h	6.6261×10^{-34}	J s
	$\hbar = h/2\pi$	1.0546×10^{-34}	J s
Speed of light in vacuum	c	2.9979×10^{8}	$\mathrm{m\,s^{-1}}$
Permittivity of free space	ϵ_0	8.8542×10^{-12}	$\mathrm{F\,m^{-1}}$
Permeability of free space	μ_0	$4\pi \times 10^{-7}$	$\mathrm{H\,m^{-1}}$
Proton/electron mass ratio	m_p/m_e	1.8362×10^{3}	
Electron charge/mass ratio	e/m_e	1.7588×10^{11}	$\mathrm{C\,kg^{-1}}$
Rydberg constant	$R_\infty = \dfrac{me^4}{8\epsilon_0{}^2 ch^3}$	1.0974×10^{7}	$\mathrm{m^{-1}}$
Bohr radius	$a_0 = \epsilon_0 h^2/\pi me^2$	5.2918×10^{-11}	m
Atomic cross section	$\pi a_0{}^2$	8.7974×10^{-21}	$\mathrm{m^2}$
Classical electron radius	$r_e = e^2/4\pi\epsilon_0 mc^2$	2.8179×10^{-15}	m
Thomson cross section	$(8\pi/3) r_e{}^2$	6.6525×10^{-29}	$\mathrm{m^2}$
Compton wavelength of electron	$h/m_e c$	2.4263×10^{-12}	m
	$\hbar/m_e c$	3.8616×10^{-13}	m
Fine-structure constant	$\alpha = e^2/2\epsilon_0 hc$	7.2974×10^{-3}	
	α^{-1}	137.04	
First radiation constant	$c_1 = 2\pi hc^2$	3.7418×10^{-16}	$\mathrm{W\,m^2}$
Second radiation constant	$c_2 = hc/k$	1.4388×10^{-2}	m K
Stefan-Boltzmann constant	σ	5.6705×10^{-8}	$\mathrm{W\,m^{-2}K^{-4}}$

Physical Quantity	Symbol	Value	Units
Wavelength associated with 1 eV	$\lambda_0 = hc/e$	1.2398×10^{-6}	m
Frequency associated with 1 eV	$\nu_0 = e/h$	2.4180×10^{14}	Hz
Wave number associated with 1 eV	$k_0 = e/hc$	8.0655×10^5	m^{-1}
Energy associated with 1 eV	$h\nu_0$	1.6022×10^{-19}	J
Energy associated with 1 m^{-1}	hc	1.9864×10^{-25}	J
Energy associated with 1 Rydberg	$me^3/8\epsilon_0{}^2 h^2$	13.606	eV
Energy associated with 1 Kelvin	k/e	8.6174×10^{-5}	eV
Temperature associated with 1 eV	e/k	1.1604×10^4	K
Avogadro number	N_A	6.0221×10^{23}	mol^{-1}
Faraday constant	$F = N_A e$	9.6485×10^4	$C\,mol^{-1}$
Gas constant	$R = N_A k$	8.3145	$J\,K^{-1}mol^{-1}$
Loschmidt's number (no. density at STP)	n_0	2.6868×10^{25}	m^{-3}
Atomic mass unit	m_u	1.6605×10^{-27}	kg
Standard temperature	T_0	273.15	K
Atmospheric pressure	$p_0 = n_0 k T_0$	1.0133×10^5	Pa
Pressure of 1 mm Hg (1 torr)		1.3332×10^2	Pa
Molar volume at STP	$V_0 = RT_0/p_0$	2.2414×10^{-2}	m^3
Molar weight of air	M_{air}	2.8971×10^{-2}	kg
calorie (cal)		4.1868	J
Gravitational acceleration	g	9.8067	$m\,s^{-2}$

PHYSICAL CONSTANTS (cgs)[7]

Physical Quantity	Symbol	Value	Units
Boltzmann constant	k	1.3807×10^{-16}	erg/deg (K)
Elementary charge	e	4.8032×10^{-10}	statcoulomb (statcoul)
Electron mass	m_e	9.1094×10^{-28}	g
Proton mass	m_p	1.6726×10^{-24}	g
Gravitational constant	G	6.6726×10^{-8}	dyne-cm^2/g^2
Planck constant	h	6.6261×10^{-27}	erg-sec
	$\hbar = h/2\pi$	1.0546×10^{-27}	erg-sec
Speed of light in vacuum	c	2.9979×10^{10}	cm/sec
Proton/electron mass ratio	m_p/m_e	1.8362×10^3	
Electron charge/mass ratio	e/m_e	5.2728×10^{17}	statcoul/g
Rydberg constant	$R_\infty = \dfrac{2\pi^2 m e^4}{ch^3}$	1.0974×10^5	cm^{-1}
Bohr radius	$a_0 = \hbar^2/me^2$	5.2918×10^{-9}	cm
Atomic cross section	$\pi a_0{}^2$	8.7974×10^{-17}	cm^2
Classical electron radius	$r_e = e^2/mc^2$	2.8179×10^{-13}	cm
Thomson cross section	$(8\pi/3)r_e{}^2$	6.6525×10^{-25}	cm^2
Compton wavelength of electron	$h/m_e c$	2.4263×10^{-10}	cm
	$\hbar/m_e c$	3.8616×10^{-11}	cm
Fine-structure constant	$\alpha = e^2/\hbar c$	7.2974×10^{-3}	
	α^{-1}	137.04	
First radiation constant	$c_1 = 2\pi \hbar c^2$	3.7418×10^{-5}	erg-cm^2/sec
Second radiation constant	$c_2 = hc/k$	1.4388	cm-deg (K)
Stefan-Boltzmann constant	σ	5.6705×10^{-5}	erg/cm^2-sec-deg^4
Wavelength associated with 1 eV	λ_0	1.2398×10^{-4}	cm

Physical Quantity	Symbol	Value	Units
Frequency associated with 1 eV	ν_0	2.4180×10^{14}	Hz
Wave number associated with 1 eV	k_0	8.0655×10^3	cm^{-1}
Energy associated with 1 eV		1.6022×10^{-12}	erg
Energy associated with 1 cm^{-1}		1.9864×10^{-16}	erg
Energy associated with 1 Rydberg		13.606	eV
Energy associated with 1 deg Kelvin		8.6174×10^{-5}	eV
Temperature associated with 1 eV		1.1604×10^4	deg (K)
Avogadro number	N_A	6.0221×10^{23}	mol^{-1}
Faraday constant	$F = N_A e$	2.8925×10^{14}	statcoul/mol
Gas constant	$R = N_A k$	8.3145×10^7	erg/deg-mol
Loschmidt's number (no. density at STP)	n_0	2.6868×10^{19}	cm^{-3}
Atomic mass unit	m_u	1.6605×10^{-24}	g
Standard temperature	T_0	273.15	deg (K)
Atmospheric pressure	$p_0 = n_0 k T_0$	1.0133×10^6	dyne/cm^2
Pressure of 1 mm Hg (1 torr)		1.3332×10^3	dyne/cm^2
Molar volume at STP	$V_0 = RT_0/p_0$	2.2414×10^4	cm^3
Molar weight of air	M_{air}	28.971	g
calorie (cal)		4.1868×10^7	erg
Gravitational acceleration	g	980.67	cm/sec^2

FORMULA CONVERSION[8]

Here $\alpha = 10^2\,\mathrm{cm\,m^{-1}}$, $\beta = 10^7\,\mathrm{erg\,J^{-1}}$, $\epsilon_0 = 8.8542 \times 10^{-12}\,\mathrm{F\,m^{-1}}$, $\mu_0 = 4\pi \times 10^{-7}\,\mathrm{H\,m^{-1}}$, $c = (\epsilon_0\mu_0)^{-1/2} = 2.9979 \times 10^8\,\mathrm{m\,s^{-1}}$, and $\hbar = 1.0546 \times 10^{-34}\,\mathrm{J\,s}$. To derive a dimensionally correct SI formula from one expressed in Gaussian units, substitute for each quantity according to $\bar{Q} = \bar{k}Q$, where $\bar{k}$ is the coefficient in the second column of the table corresponding to Q (overbars denote variables expressed in Gaussian units). Thus, the formula $\bar{a}_0 = \bar{\hbar}^2/\bar{m}\bar{e}^2$ for the Bohr radius becomes $\alpha a_0 = (\hbar\beta)^2/[(m\beta/\alpha^2)(e^2\alpha\beta/4\pi\epsilon_0)]$, or $a_0 = \epsilon_0 h^2/\pi m e^2$. To go from SI to natural units in which $\hbar = c = 1$ (distinguished by a circumflex), use $Q = \hat{k}^{-1}\hat{Q}$, where $\hat{k}$ is the coefficient corresponding to Q in the third column. Thus $\hat{a}_0 = 4\pi\epsilon_0\hbar^2/[(\hat{m}\hbar/c)(\hat{e}^2\epsilon_0\hbar c)] = 4\pi/\hat{m}\hat{e}^2$. (In transforming *from* SI units, do not substitute for ϵ_0, μ_0, or c.)

Physical Quantity	Gaussian Units to SI	Natural Units to SI
Capacitance	$\alpha/4\pi\epsilon_0$	ϵ_0^{-1}
Charge	$(\alpha\beta/4\pi\epsilon_0)^{1/2}$	$(\epsilon_0\hbar c)^{-1/2}$
Charge density	$(\beta/4\pi\alpha^5\epsilon_0)^{1/2}$	$(\epsilon_0\hbar c)^{-1/2}$
Current	$(\alpha\beta/4\pi\epsilon_0)^{1/2}$	$(\mu_0/\hbar c)^{1/2}$
Current density	$(\beta/4\pi\alpha^3\epsilon_0)^{1/2}$	$(\mu_0/\hbar c)^{1/2}$
Electric field	$(4\pi\beta\epsilon_0/\alpha^3)^{1/2}$	$(\epsilon_0/\hbar c)^{1/2}$
Electric potential	$(4\pi\beta\epsilon_0/\alpha)^{1/2}$	$(\epsilon_0/\hbar c)^{1/2}$
Electric conductivity	$(4\pi\epsilon_0)^{-1}$	ϵ_0^{-1}
Energy	β	$(\hbar c)^{-1}$
Energy density	β/α^3	$(\hbar c)^{-1}$
Force	β/α	$(\hbar c)^{-1}$
Frequency	1	c^{-1}
Inductance	$4\pi\epsilon_0/\alpha$	μ_0^{-1}
Length	α	1
Magnetic induction	$(4\pi\beta/\alpha^3\mu_0)^{1/2}$	$(\mu_0\hbar c)^{-1/2}$
Magnetic intensity	$(4\pi\mu_0\beta/\alpha^3)^{1/2}$	$(\mu_0/\hbar c)^{1/2}$
Mass	β/α^2	$c/\hbar$
Momentum	β/α	$\hbar^{-1}$
Power	β	$(\hbar c^2)^{-1}$
Pressure	β/α^3	$(\hbar c)^{-1}$
Resistance	$4\pi\epsilon_0/\alpha$	$(\epsilon_0/\mu_0)^{1/2}$
Time	1	c
Velocity	α	c^{-1}

MAXWELL'S EQUATIONS

Name or Description	SI	Gaussian
Faraday's law	$\nabla \times \mathbf{E} = -\dfrac{\partial \mathbf{B}}{\partial t}$	$\nabla \times \mathbf{E} = -\dfrac{1}{c}\dfrac{\partial \mathbf{B}}{\partial t}$
Ampere's law	$\nabla \times \mathbf{H} = \dfrac{\partial \mathbf{D}}{\partial t} + \mathbf{J}$	$\nabla \times \mathbf{H} = \dfrac{1}{c}\dfrac{\partial \mathbf{D}}{\partial t} + \dfrac{4\pi}{c}\mathbf{J}$
Poisson equation	$\nabla \cdot \mathbf{D} = \rho$	$\nabla \cdot \mathbf{D} = 4\pi\rho$
[Absence of magnetic monopoles]	$\nabla \cdot \mathbf{B} = 0$	$\nabla \cdot \mathbf{B} = 0$
Lorentz force on charge q	$q\left(\mathbf{E} + \mathbf{v} \times \mathbf{B}\right)$	$q\left(\mathbf{E} + \dfrac{1}{c}\mathbf{v} \times \mathbf{B}\right)$
Constitutive relations	$\mathbf{D} = \epsilon\mathbf{E}$ $\mathbf{B} = \mu\mathbf{H}$	$\mathbf{D} = \epsilon\mathbf{E}$ $\mathbf{B} = \mu\mathbf{H}$

In a plasma, $\mu \approx \mu_0 = 4\pi \times 10^{-7}\,\mathrm{H\,m^{-1}}$ (Gaussian units: $\mu \approx 1$). The permittivity satisfies $\epsilon \approx \epsilon_0 = 8.8542 \times 10^{-12}\,\mathrm{F\,m^{-1}}$ (Gaussian: $\epsilon \approx 1$) provided that all charge is regarded as free. Using the drift approximation $\mathbf{v}_\perp = \mathbf{E} \times \mathbf{B}/B^2$ to calculate polarization charge density gives rise to a dielectric constant $K \equiv \epsilon/\epsilon_0 = 1 + 36\pi \times 10^9 \rho/B^2$ (SI) $= 1 + 4\pi\rho c^2/B^2$ (Gaussian), where ρ is the mass density.

The electromagnetic energy in volume V is given by

$$W = \frac{1}{2}\int_V dV(\mathbf{H} \cdot \mathbf{B} + \mathbf{E} \cdot \mathbf{D}) \qquad \text{(SI)}$$

$$= \frac{1}{8\pi}\int_V dV(\mathbf{H} \cdot \mathbf{B} + \mathbf{E} \cdot \mathbf{D}) \qquad \text{(Gaussian).}$$

Poynting's theorem is

$$\frac{\partial W}{\partial t} + \int_S \mathbf{N} \cdot d\mathbf{S} = -\int_V dV \mathbf{J} \cdot \mathbf{E},$$

where S is the closed surface bounding V and the Poynting vector (energy flux across S) is given by $\mathbf{N} = \mathbf{E} \times \mathbf{H}$ (SI) or $\mathbf{N} = c\mathbf{E} \times \mathbf{H}/4\pi$ (Gaussian).

ELECTRICITY AND MAGNETISM

In the following, ϵ = dielectric permittivity, μ = permeability of conductor, μ' = permeability of surrounding medium, σ = conductivity, $f = \omega/2\pi$ = radiation frequency, $\kappa_m = \mu/\mu_0$ and $\kappa_e = \epsilon/\epsilon_0$. Where subscripts are used, '1' denotes a conducting medium and '2' a propagating (lossless dielectric) medium. All units are SI unless otherwise specified.

Permittivity of free space	$\epsilon_0 = 8.8542 \times 10^{-12}\,\mathrm{F\,m^{-1}}$
Permeability of free space	$\mu_0 = 4\pi \times 10^{-7}\,\mathrm{H\,m^{-1}}$
	$ = 1.2566 \times 10^{-6}\,\mathrm{H\,m^{-1}}$
Resistance of free space	$R_0 = (\mu_0/\epsilon_0)^{1/2} = 376.73\,\Omega$
Capacity of parallel plates of area A, separated by distance d	$C = \epsilon A/d$
Capacity of concentric cylinders of length l, radii a, b	$C = 2\pi\epsilon l / \ln(b/a)$
Capacity of concentric spheres of radii a, b	$C = 4\pi\epsilon ab/(b-a)$
Self-inductance of wire of length l, carrying uniform current	$L = \mu l$
Mutual inductance of parallel wires of length l, radius a, separated by distance d	$L = (\mu'l/4\pi)\,[1 + 4\ln(d/a)]$
Inductance of circular loop of radius b, made of wire of radius a, carrying uniform current	$L = b\left\{ \mu'\,[\ln(8b/a) - 2] + \mu/4 \right\}$
Relaxation time in a lossy medium	$\tau = \epsilon/\sigma$
Skin depth in a lossy medium	$\delta = (2/\omega\mu\sigma)^{1/2} = (\pi f\mu\sigma)^{-1/2}$
Wave impedance in a lossy medium	$Z = [\mu/(\epsilon + i\sigma/\omega)]^{1/2}$
Transmission coefficient at conducting surface[9] (good only for $T \ll 1$)	$T = 4.22 \times 10^{-4}(f\kappa_{m1}\kappa_{e2}/\sigma)^{1/2}$
Field at distance r from straight wire carrying current I (amperes)	$B_\theta = \mu I/2\pi r$ tesla
	$ = 0.2I/r$ gauss (r in cm)
Field at distance z along axis from circular loop of radius a carrying current I	$B_z = \mu a^2 I/[2(a^2 + z^2)^{3/2}]$

ELECTROMAGNETIC FREQUENCY/ WAVELENGTH BANDS[10]

Designation	Frequency Range		Wavelength Range	
	Lower	Upper	Lower	Upper
ULF*		30 Hz	10 Mm	
VF*	30 Hz	300 Hz	1 Mm	10 Mm
ELF	300 Hz	3 kHz	100 km	1 Mm
VLF	3 kHz	30 kHz	10 km	100 km
LF	30 kHz	300 kHz	1 km	10 km
MF	300 kHz	3 MHz	100 m	1 km
HF	3 MHz	30 MHz	10 m	100 m
VHF	30 MHz	300 MHz	1 m	10 m
UHF	300 MHz	3 GHz	10 cm	1 m
SHF†	3 GHz	30 GHz	1 cm	10 cm
S	2.6	3.95	7.6	11.5
G	3.95	5.85	5.1	7.6
J	5.3	8.2	3.7	5.7
H	7.05	10.0	3.0	4.25
X	8.2	12.4	2.4	3.7
M	10.0	15.0	2.0	3.0
P	12.4	18.0	1.67	2.4
K	18.0	26.5	1.1	1.67
R	26.5	40.0	0.75	1.1
EHF	30 GHz	300 GHz	1 mm	1 cm
Submillimeter	300 GHz	3 THz	100 μm	1 mm
Infrared	3 THz	430 THz	700 nm	100 μm
Visible	430 THz	750 THz	400 nm	700 nm
Ultraviolet	750 THz	30 PHz	10 nm	400 nm
X Ray	30 PHz	3 EHz	100 pm	10 nm
Gamma Ray	3 EHz			100 pm

In spectroscopy the angstrom is sometimes used ($1\text{Å} = 10^{-8}$ cm $= 0.1$ nm).

*The boundary between ULF and VF (voice frequencies) is variously defined.

†The SHF (microwave) band is further subdivided approximately as shown.[11]

AC CIRCUITS

For a resistance R, inductance L, and capacitance C in series with a voltage source $V = V_0 \exp(i\omega t)$ (here $i = \sqrt{-1}$), the current is given by $I = dq/dt$, where q satisfies

$$L\frac{d^2q}{dt^2} + R\frac{dq}{dt} + \frac{q}{C} = V.$$

Solutions are $q(t) = q_s + q_t$, $I(t) = I_s + I_t$, where the steady state is $I_s = i\omega q_s = V/Z$ in terms of the impedance $Z = R + i(\omega L - 1/\omega C)$ and $I_t = dq_t/dt$. For initial conditions $q(0) \equiv q_0 = \bar{q}_0 + q_s$, $I(0) \equiv I_0$, the transients can be of three types, depending on $\Delta = R^2 - 4L/C$:

(a) Overdamped, $\Delta > 0$

$$q_t = \frac{I_0 + \gamma_+\bar{q}_0}{\gamma_+ - \gamma_-}\exp(-\gamma_- t) - \frac{I_0 + \gamma_-\bar{q}_0}{\gamma_+ - \gamma_-}\exp(-\gamma_+ t),$$

$$I_t = \frac{\gamma_+(I_0 + \gamma_-\bar{q}_0)}{\gamma_+ - \gamma_-}\exp(-\gamma_+ t) - \frac{\gamma_-(I_0 + \gamma_+\bar{q}_0)}{\gamma_+ - \gamma_-}\exp(-\gamma_- t),$$

where $\gamma_\pm = (R \pm \Delta^{1/2})/2L$;

(b) Critically damped, $\Delta = 0$

$$q_t = [\bar{q}_0 + (I_0 + \gamma_R\bar{q}_0)t]\exp(-\gamma_R t),$$

$$I_t = [I_0 - (I_0 + \gamma_R\bar{q}_0)\gamma_R t]\exp(-\gamma_R t),$$

where $\gamma_R = R/2L$;

(c) Underdamped, $\Delta < 0$

$$q_t = \left[\frac{\gamma_R\bar{q}_0 + I_0}{\omega_1}\sin\omega_1 t + \bar{q}_0\cos\omega_1 t\right]\exp(-\gamma_R t),$$

$$I_t = \left[I_0\cos\omega_1 t - \frac{(\omega_1^2 + \gamma_R^2)\bar{q}_0 + \gamma_R I_0}{\omega_1}\sin(\omega_1 t)\right]\exp(-\gamma_R t),$$

Here $\omega_1 = \omega_0(1 - R^2C/4L)^{1/2}$, where $\omega_0 = (LC)^{-1/2}$ is the resonant frequency. At $\omega = \omega_0$, $Z = R$. The quality of the circuit is $Q = \omega_0 L/R$. Instability results when L, R, C are not all of the same sign.

DIMENSIONLESS NUMBERS OF FLUID MECHANICS[12]

Name(s)	Symbol	Definition	Significance
Alfvén, Kármán	Al, Ka	V_A/V	*(Magnetic force/ inertial force)$^{1/2}$
Bond	Bd	$(\rho' - \rho)L^2 g/\Sigma$	Gravitational force/ surface tension
Boussinesq	B	$V/(2gR)^{1/2}$	(Inertial force/ gravitational force)$^{1/2}$
Brinkman	Br	$\mu V^2/k\Delta T$	Viscous heat/conducted heat
Capillary	Cp	$\mu V/\Sigma$	Viscous force/surface tension
Carnot	Ca	$(T_2 - T_1)/T_2$	Theoretical Carnot cycle efficiency
Cauchy, Hooke	Cy, Hk	$\rho V^2/\Gamma = \mathrm{M}^2$	Inertial force/ compressibility force
Chandra-sekhar	Ch	$B^2 L^2/\rho\nu\eta$	Magnetic force/dissipative forces
Clausius	Cl	$LV^3\rho/k\Delta T$	Kinetic energy flow rate/heat conduction rate
Cowling	C	$(V_A/V)^2 = \mathrm{Al}^2$	Magnetic force/inertial force
Crispation	Cr	$\mu\kappa/\Sigma L$	Effect of diffusion/effect of surface tension
Dean	D	$D^{3/2}V/\nu(2r)^{1/2}$	Transverse flow due to curvature/longitudinal flow
[Drag coefficient]	C_D	$(\rho' - \rho)Lg/ \rho' V^2$	Drag force/inertial force
Eckert	E	$V^2/c_p\Delta T$	Kinetic energy/change in thermal energy
Ekman	Ek	$(\nu/2\Omega L^2)^{1/2} = (\mathrm{Ro/Re})^{1/2}$	(Viscous force/Coriolis force)$^{1/2}$
Euler	Eu	$\Delta p/\rho V^2$	Pressure drop due to friction/ dynamic pressure
Froude	Fr	$V/(gL)^{1/2}$ V/NL	†(Inertial force/gravitational or buoyancy force)$^{1/2}$
Gay–Lussac	Ga	$1/\beta\Delta T$	Inverse of relative change in volume during heating
Grashof	Gr	$gL^3\beta\Delta T/\nu^2$	Buoyancy force/viscous force
[Hall coefficient]	C_H	λ/r_L	Gyrofrequency/ collision frequency

*(†) Also defined as the inverse (square) of the quantity shown.

Name(s)	Symbol	Definition	Significance
Hartmann	H	$BL/(\mu\eta)^{1/2} =$ $(\text{Rm Re C})^{1/2}$	(Magnetic force/ dissipative force)$^{1/2}$
Knudsen	Kn	λ/L	Hydrodynamic time/ collision time
Lewis	Le	$\kappa/\mathcal{D}$	*Thermal conduction/molecular diffusion
Lorentz	Lo	V/c	Magnitude of relativistic effects
Lundquist	Lu	$\mu_0 L V_A/\eta =$ Al Rm	$\mathbf{J} \times \mathbf{B}$ force/resistive magnetic diffusion force
Mach	M	V/C_S	Magnitude of compressibility effects
Magnetic Mach	Mm	$V/V_A = \text{Al}^{-1}$	(Inertial force/magnetic force)$^{1/2}$
Magnetic Reynolds	Rm	$\mu_0 L V/\eta$	Flow velocity/magnetic diffusion velocity
Newton	Nt	$F/\rho L^2 V^2$	Imposed force/inertial force
Nusselt	N	$\alpha L/k$	Total heat transfer/thermal conduction
Péclet	Pe	LV/κ	Heat convection/heat conduction
Poisseuille	Po	$D^2 \Delta p/\mu LV$	Pressure force/viscous force
Prandtl	Pr	ν/κ	Momentum diffusion/ heat diffusion
Rayleigh	Ra	$gH^3\beta\Delta T/\nu\kappa$	Buoyancy force/diffusion force
Reynolds	Re	LV/ν	Inertial force/viscous force
Richardson	Ri	$(NH/\Delta V)^2$	Buoyancy effects/ vertical shear effects
Rossby	Ro	$V/2\Omega L \sin \Lambda$	Inertial force/Coriolis force
Schmidt	Sc	$\nu/\mathcal{D}$	Momentum diffusion/ molecular diffusion
Stanton	St	$\alpha/\rho c_p V$	Thermal conduction loss/ heat capacity
Stefan	Sf	$\sigma L T^3/k$	Radiated heat/conducted heat
Stokes	S	$\nu/L^2 f$	Viscous damping rate/ vibration frequency
Strouhal	Sr	fL/V	Vibration speed/flow velocity
Taylor	Ta	$(2\Omega L^2/\nu)^2$ $R^{1/2}(\Delta R)^{3/2}$ $\cdot(\Omega/\nu)$	Centrifugal force/viscous force (Centrifugal force/ viscous force)$^{1/2}$
Thring, Boltzmann	Th, Bo	$\rho c_p V/\epsilon\sigma T^3$	Convective heat transport/ radiative heat transport
Weber	W	$\rho L V^2/\Sigma$	Inertial force/surface tension

Nomenclature:

B	Magnetic induction
C_s, c	Speeds of sound, light
c_p	Specific heat at constant pressure (units $\mathrm{m^2\,s^{-2}\,K^{-1}}$)
$D = 2R$	Pipe diameter
F	Imposed force
f	Vibration frequency
g	Gravitational acceleration
H, L	Vertical, horizontal length scales
$k = \rho c_p \kappa$	Thermal conductivity (units $\mathrm{kg\,m^{-1}\,s^{-2}}$)
$N = (g/H)^{1/2}$	Brunt–Väisälä frequency
R	Radius of pipe or channel
r	Radius of curvature of pipe or channel
r_L	Larmor radius
T	Temperature
V	Characteristic flow velocity
$V_A = B/(\mu_0 \rho)^{1/2}$	Alfvén speed
α	Newton's-law heat coefficient, $k\dfrac{\partial T}{\partial x} = \alpha \Delta T$
β	Volumetric expansion coefficient, $dV/V = \beta dT$
Γ	Bulk modulus (units $\mathrm{kg\,m^{-1}\,s^{-2}}$)
$\Delta R, \Delta V, \Delta p, \Delta T$	Imposed differences in two radii, velocities, pressures, or temperatures
ϵ	Surface emissivity
η	Electrical resistivity
$\kappa, \mathcal{D}$	Thermal, molecular diffusivities (units $\mathrm{m^2\,s^{-1}}$)
Λ	Latitude of point on earth's surface
λ	Collisional mean free path
$\mu = \rho \nu$	Viscosity
μ_0	Permeability of free space
ν	Kinematic viscosity (units $\mathrm{m^2\,s^{-1}}$)
ρ	Mass density of fluid medium
ρ'	Mass density of bubble, droplet, or moving object
Σ	Surface tension (units $\mathrm{kg\,s^{-2}}$)
σ	Stefan–Boltzmann constant
Ω	Solid-body rotational angular velocity

SHOCKS

At a shock front propagating in a magnetized fluid at an angle θ with respect to the magnetic induction $\mathbf{B}$, the jump conditions are [13,14]

(1) $\rho U = \bar{\rho}\bar{U} \equiv q$;

(2) $\rho U^2 + p + B_\perp^2/2\mu = \bar{\rho}\bar{U}^2 + \bar{p} + \bar{B}_\perp^2/2\mu$;

(3) $\rho U V - B_\parallel B_\perp/\mu = \bar{\rho}\bar{U}\bar{V} - \bar{B}_\parallel \bar{B}_\perp/\mu$;

(4) $B_\parallel = \bar{B}_\parallel$;

(5) $U B_\perp - V B_\parallel = \bar{U}\bar{B}_\perp - \bar{V}\bar{B}_\parallel$;

(6) $\frac{1}{2}(U^2 + V^2) + w + (U B_\perp^2 - V B_\parallel B_\perp)/\mu\rho U$
$\qquad = \frac{1}{2}(\bar{U}^2 + \bar{V}^2) + \bar{w} + (\bar{U}\bar{B}_\perp^2 - \bar{V}\bar{B}_\parallel \bar{B}_\perp)/\mu\bar{\rho}\bar{U}$.

Here U and V are components of the fluid velocity normal and tangential to the front in the shock frame; $\rho = 1/v$ is the mass density; p is the pressure; $B_\perp = B\sin\theta$, $B_\parallel = B\cos\theta$; μ is the magnetic permeability ($\mu = 4\pi$ in cgs units); and the specific enthalpy is $w = e + pv$, where the specific internal energy e satisfies $de = T ds - p dv$ in terms of the temperature T and the specific entropy s. Quantities in the region behind (downstream from) the front are distinguished by a bar. If $\mathbf{B} = 0$, then [15]

(7) $U - \bar{U} = [(\bar{p} - p)(v - \bar{v})]^{1/2}$;

(8) $(\bar{p} - p)(v - \bar{v})^{-1} = q^2$;

(9) $\bar{w} - w = \frac{1}{2}(\bar{p} - p)(v - \bar{v})$;

(10) $\bar{e} - e = \frac{1}{2}(\bar{p} + p)(v - \bar{v})$.

In what follows we assume that the fluid is a perfect gas with adiabatic index $\gamma = 1 + 2/n$, where n is the number of degrees of freedom. Then $p = \rho RT/m$, where R is the universal gas constant and m is the molar weight; the sound speed is given by $C_s^2 = (\partial_p/\partial\rho)_s = \gamma pv$; and $w = \gamma e = \gamma pv/(\gamma - 1)$. For a general oblique shock in a perfect gas the quantity $X = r^{-1}(U/V_A)^2$ satisfies [14]

(11) $(X - \beta/\alpha)(X - \cos^2\theta)^2 = X\sin^2\theta\left\{[1 + (r-1)/2\alpha]X - \cos^2\theta\right\}$, where
$r = \bar{\rho}/\rho$, $\alpha = \frac{1}{2}[\gamma + 1 - (\gamma - 1)r]$, and $\beta = C_s^2/V_A^2 = 4\pi\gamma p/B^2$.

The density ratio is bounded by

(12) $1 < r < (\gamma + 1)/(\gamma - 1)$.

If the shock is normal to $\mathbf{B}$ (i.e., if $\theta = \pi/2$), then

(13) $U^2 = (r/\alpha)\left\{C_s^2 + V_A^2[1 + (1 - \gamma/2)(r - 1)]\right\}$;

(14) $U/\bar{U} = \bar{B}/B = r$;

(15) $\bar{V} = V$;

(16) $\bar{p} = p + (1 - r^{-1})\rho U^2 + (1 - r^2)B^2/2\mu$.

If $\theta = 0$, there are two possibilities: switch-on shocks, which require $\beta < 1$ and for which

(17) $U^2 = rV_A{}^2$;

(18) $\bar{U} \doteq V_A{}^2/U$;

(19) $\bar{B}_\perp{}^2 = 2B_\parallel{}^2(r - 1)(\alpha - \beta)$;

(20) $\bar{V} = \bar{U}\bar{B}_\perp/B_\parallel$;

(21) $\bar{p} = p + \rho U^2(1 - \alpha + \beta)(1 - r^{-1})$,

and acoustic (hydrodynamic) shocks, for which

(22) $U^2 = (r/\alpha)C_s{}^2$;

(23) $\bar{U} = U/r$;

(24) $\bar{V} = \bar{B}_\perp = 0$;

(25) $\bar{p} = p + \rho U^2(1 - r^{-1})$.

For acoustic shocks the specific volume and pressure are related by

(26) $\bar{v}/v = [(\gamma + 1)p + (\gamma - 1)\bar{p}] / [(\gamma - 1)p + (\gamma + 1)\bar{p}]$.

In terms of the upstream Mach number $M = U/C_s$,

(27) $\bar{\rho}/\rho = v/\bar{v} = U/\bar{U} = (\gamma + 1)M^2/[(\gamma - 1)M^2 + 2]$;

(28) $\bar{p}/p = (2\gamma M^2 - \gamma + 1)/(\gamma + 1)$;

(29) $\bar{T}/T = [(\gamma - 1)M^2 + 2](2\gamma M^2 - \gamma + 1)/(\gamma + 1)^2 M^2$;

(30) $\bar{M}^2 = [(\gamma - 1)M^2 + 2]/[2\gamma M^2 - \gamma + 1]$.

The entropy change across the shock is

(31) $\Delta s \equiv \bar{s} - s = c_v \ln[(\bar{p}/p)(\rho/\bar{\rho})^\gamma]$,

where $c_v = R/(\gamma - 1)m$ is the specific heat at constant volume; here R is the gas constant. In the weak-shock limit ($M \to 1$),

(32) $\Delta s \to c_v \dfrac{2\gamma(\gamma - 1)}{3(\gamma + 1)}(M^2 - 1)^3 \approx \dfrac{16\gamma R}{3(\gamma + 1)m}(M - 1)^3$.

The radius at time t of a strong spherical blast wave resulting from the explosive release of energy E in a medium with uniform density ρ is

(33) $R_S = C_0(Et^2/\rho)^{1/5}$,

where C_0 is a constant depending on γ. For $\gamma = 7/5$, $C_0 = 1.033$.

27

FUNDAMENTAL PLASMA PARAMETERS

All quantities are in Gaussian cgs units except temperature (T, T_e, T_i) expressed in eV and ion mass (m_i) expressed in units of the proton mass, $\mu = m_i/m_p$; Z is charge state; k is Boltzmann's constant; K is wavenumber; γ is the adiabatic index; $\ln \Lambda$ is the Coulomb logarithm.

Frequencies

electron gyrofrequency

$$f_{ce} = \omega_{ce}/2\pi = 2.80 \times 10^6 B \, \text{Hz}$$
$$\omega_{ce} = eB/m_e c = 1.76 \times 10^7 B \, \text{rad/sec}$$

ion gyrofrequency

$$f_{ci} = \omega_{ci}/2\pi = 1.52 \times 10^3 Z\mu^{-1}B \, \text{Hz}$$
$$\omega_{ci} = ZeB/m_i c = 9.58 \times 10^3 Z\mu^{-1}B \, \text{rad/sec}$$

electron plasma frequency

$$f_{pe} = \omega_{pe}/2\pi = 8.98 \times 10^3 n_e{}^{1/2} \, \text{Hz}$$
$$\omega_{pe} = (4\pi n_e e^2/m_e)^{1/2}$$
$$= 5.64 \times 10^4 n_e{}^{1/2} \, \text{rad/sec}$$

ion plasma frequency

$$f_{pi} = \omega_{pi}/2\pi$$
$$= 2.10 \times 10^2 Z\mu^{-1/2}n_i{}^{1/2} \, \text{Hz}$$
$$\omega_{pi} = (4\pi n_i Z^2 e^2/m_i)^{1/2}$$
$$= 1.32 \times 10^3 Z\mu^{-1/2}n_i{}^{1/2} \text{rad/sec}$$

electron trapping rate

$$\nu_{Te} = (eKE/m_e)^{1/2}$$
$$= 7.26 \times 10^8 K^{1/2}E^{1/2} \, \text{sec}^{-1}$$

ion trapping rate

$$\nu_{Ti} = (ZeKE/m_i)^{1/2}$$
$$= 1.69 \times 10^7 Z^{1/2}K^{1/2}E^{1/2}\mu^{-1/2} \, \text{sec}^{-1}$$

electron collision rate

$$\nu_e = 2.91 \times 10^{-6} n_e \ln \Lambda T_e{}^{-3/2} \, \text{sec}^{-1}$$

ion collision rate

$$\nu_i = 4.80 \times 10^{-8} Z^4 \mu^{-1/2} n_i \ln \Lambda T_i{}^{-3/2} \, \text{sec}^{-1}$$

Lengths

electron deBroglie length

$$\lambda = \hbar/(m_e kT_e)^{1/2} = 2.76 \times 10^{-8} T_e{}^{-1/2} \, \text{cm}$$

classical distance of minimum approach

$$e^2/kT = 1.44 \times 10^{-7} T^{-1} \, \text{cm}$$

electron gyroradius

$$r_e = v_{Te}/\omega_{ce} = 2.38 T_e{}^{1/2} B^{-1} \, \text{cm}$$

ion gyroradius

$$r_i = v_{Ti}/\omega_{ci}$$
$$= 1.02 \times 10^2 \mu^{1/2} Z^{-1} T_i{}^{1/2} B^{-1} \, \text{cm}$$

electron inertial length

$$c/\omega_{pe} = 5.31 \times 10^5 n_e{}^{-1/2} \, \text{cm}$$

ion inertial length

$$c/\omega_{pi} = 2.28 \times 10^7 Z^{-1}(\mu/n_i)^{1/2} \, \text{cm}$$

Debye length

$$\lambda_D = (kT/4\pi n e^2)^{1/2} = 7.43 \times 10^2 T^{1/2} n^{-1/2}$$

Velocities

electron thermal velocity
$$v_{Te} = (kT_e/m_e)^{1/2}$$
$$= 4.19 \times 10^7 T_e{}^{1/2} \text{ cm/sec}$$

ion thermal velocity
$$v_{Ti} = (kT_i/m_i)^{1/2}$$
$$= 9.79 \times 10^5 \mu^{-1/2} T_i{}^{1/2} \text{ cm/sec}$$

ion sound velocity
$$C_s = (\gamma Z kT_e/m_i)^{1/2}$$
$$= 9.79 \times 10^5 (\gamma Z T_e/\mu)^{1/2} \text{ cm/sec}$$

Alfvén velocity
$$v_A = B/(4\pi n_i m_i)^{1/2}$$
$$= 2.18 \times 10^{11} \mu^{-1/2} n_i{}^{-1/2} B \text{ cm/sec}$$

Dimensionless

(electron/proton mass ratio)$^{1/2}$
$$(m_e/m_p)^{1/2} = 2.33 \times 10^{-2} = 1/42.9$$

number of particles in Debye sphere
$$(4\pi/3)n\lambda_D{}^3 = 1.72 \times 10^9 T^{3/2} n^{-1/2}$$

Alfvén velocity/speed of light
$$v_A/c = 7.28 \mu^{-1/2} n_i{}^{-1/2} B$$

electron plasma/gyrofrequency ratio
$$\omega_{pe}/\omega_{ce} = 3.21 \times 10^{-3} n_e{}^{1/2} B^{-1}$$

ion plasma/gyrofrequency ratio
$$\omega_{pi}/\omega_{ci} = 0.137 \mu^{1/2} n_i{}^{1/2} B^{-1}$$

thermal/magnetic energy ratio
$$\beta = 8\pi nkT/B^2 = 4.03 \times 10^{-11} nTB^{-2}$$

magnetic/ion rest energy ratio
$$B^2/8\pi n_i m_i c^2 = 26.5 \mu^{-1} n_i{}^{-1} B^2$$

Miscellaneous

Bohm diffusion coefficient
$$D_B = (ckT/16eB)$$
$$= 6.25 \times 10^6 TB^{-1} \text{ cm}^2/\text{sec}$$

transverse Spitzer resistivity
$$\eta_\perp = 1.15 \times 10^{-14} Z \ln \Lambda T^{-3/2} \text{ sec}$$
$$= 1.03 \times 10^{-2} Z \ln \Lambda T^{-3/2} \,\Omega\,\text{cm}$$

The anomalous collision rate due to low-frequency ion-sound turbulence is

$$\nu^* \approx \omega_{pe} \widetilde{W}/kT = 5.64 \times 10^4 n_e{}^{1/2} \widetilde{W}/kT \text{ sec}^{-1},$$

where $\widetilde{W}$ is the total energy of waves with $\omega/K < v_{Ti}$.
Magnetic pressure is given by

$$P_{\text{mag}} = B^2/8\pi = 3.98 \times 10^6 (B/B_0)^2 \text{ dynes/cm}^2 = 3.93(B/B_0)^2 \text{ atm},$$

where $B_0 = 10\,\text{kG} = 1\,\text{T}$.
Detonation energy of 1 kiloton of high explosive is

$$W_{\text{kT}} = 10^{12} \text{ cal} = 4.2 \times 10^{19} \text{ erg}.$$

PLASMA DISPERSION FUNCTION

Definition[16] (first form valid only for $\text{Im}\,\zeta > 0$):

$$Z(\zeta) = \pi^{-1/2} \int_{-\infty}^{+\infty} \frac{dt \, \exp\left(-t^2\right)}{t - \zeta} = 2i \exp\left(-\zeta^2\right) \int_{-\infty}^{i\zeta} dt \, \exp\left(-t^2\right).$$

Physically, $\zeta = x + iy$ is the ratio of wave phase velocity to thermal velocity. Differential equation:

$$\frac{dZ}{d\zeta} = -2\left(1 + \zeta Z\right), \quad Z(0) = i\pi^{1/2}; \quad \frac{d^2 Z}{d\zeta^2} + 2\zeta \frac{dZ}{d\zeta} + 2Z = 0.$$

Real argument ($y = 0$):

$$Z(x) = \exp\left(-x^2\right) \left(i\pi^{1/2} - 2 \int_0^x dt \, \exp\left(t^2\right)\right).$$

Imaginary argument ($x = 0$):

$$Z(iy) = i\pi^{1/2} \exp\left(y^2\right) \left[1 - \text{erf}(y)\right].$$

Power series (small argument):

$$Z(\zeta) = i\pi^{1/2} \exp\left(-\zeta^2\right) - 2\zeta\left(1 - 2\zeta^2/3 + 4\zeta^4/15 - 8\zeta^6/105 + \cdots\right).$$

Asymptotic series, $|\zeta| \gg 1$ (Ref. 17):

$$Z(\zeta) = i\pi^{1/2}\sigma \exp\left(-\zeta^2\right) - \zeta^{-1}\left(1 + 1/2\zeta^2 + 3/4\zeta^4 + 15/8\zeta^6 + \cdots\right),$$

where

$$\sigma = \begin{cases} 0 & y > |x|^{-1} \\ 1 & |y| < |x|^{-1} \\ 2 & y < -|x|^{-1} \end{cases}$$

Symmetry properties (the asterisk denotes complex conjugation):

$$Z(\zeta^*) = -\left[Z(-\zeta)\right]^*;$$

$$Z(\zeta^*) = \left[Z(\zeta)\right]^* + 2i\pi^{1/2}\exp[-(\zeta^*)^2] \quad (y > 0).$$

Two-pole approximations[18] (good for ζ in upper half plane except when $y < \pi^{1/2}x^2 \exp(-x^2)$, $x \gg 1$):

$$Z(\zeta) \approx \frac{0.50 + 0.81i}{a - \zeta} - \frac{0.50 - 0.81i}{a^* + \zeta}, \quad a = 0.51 - 0.81i;$$

$$Z'(\zeta) \approx \frac{0.50 + 0.96i}{(b - \zeta)^2} + \frac{0.50 - 0.96i}{(b^* + \zeta)^2}, \quad b = 0.48 - 0.91i.$$

COLLISIONS AND TRANSPORT

Temperatures are in eV; the corresponding value of Boltzmann's constant is $k = 1.60 \times 10^{-12}$ erg/eV; masses μ, μ' are in units of the proton mass; $e_\alpha = Z_\alpha e$ is the charge of species α. All other units are cgs except where noted.

Relaxation Rates

Rates are associated with four relaxation processes arising from the interaction of test particles (labeled α) streaming with velocity $\mathbf{v}_\alpha$ through a background of field particles (labeled β):

slowing down
$$\frac{d\mathbf{v}_\alpha}{dt} = -\nu_s^{\alpha \backslash \beta} \mathbf{v}_\alpha$$

transverse diffusion
$$\frac{d}{dt}(\mathbf{v}_\alpha - \bar{\mathbf{v}}_\alpha)_\perp^2 = \nu_\perp^{\alpha \backslash \beta} v_\alpha^2$$

parallel diffusion
$$\frac{d}{dt}(\mathbf{v}_\alpha - \bar{\mathbf{v}}_\alpha)_\parallel^2 = \nu_\parallel^{\alpha \backslash \beta} v_\alpha^2$$

energy loss
$$\frac{d}{dt} v_\alpha^2 = -\nu_\epsilon^{\alpha \backslash \beta} v_\alpha^2,$$

where $v_\alpha = |\mathbf{v}_\alpha|$ and the averages are performed over an ensemble of test particles and a Maxwellian field particle distribution. The exact formulas may be written[19]

$$\nu_s^{\alpha \backslash \beta} = (1 + m_\alpha/m_\beta)\psi(x^{\alpha \backslash \beta})\nu_0^{\alpha \backslash \beta};$$

$$\nu_\perp^{\alpha \backslash \beta} = 2\left[(1 - 1/2x^{\alpha \backslash \beta})\psi(x^{\alpha \backslash \beta}) + \psi'(x^{\alpha \backslash \beta})\right]\nu_0^{\alpha \backslash \beta};$$

$$\nu_\parallel^{\alpha \backslash \beta} = \left[\psi(x^{\alpha \backslash \beta})/x^{\alpha \backslash \beta}\right]\nu_0^{\alpha \backslash \beta};$$

$$\nu_\epsilon^{\alpha \backslash \beta} = 2\left[(m_\alpha/m_\beta)\psi(x^{\alpha \backslash \beta}) - \psi'(x^{\alpha \backslash \beta})\right]\nu_0^{\alpha \backslash \beta},$$

where

$$\nu_0^{\alpha \backslash \beta} = 4\pi e_\alpha^2 e_\beta^2 \lambda_{\alpha\beta} n_\beta / m_\alpha^2 v_\alpha^3; \qquad x^{\alpha \backslash \beta} = m_\beta v_\alpha^2 / 2kT_\beta;$$

$$\psi(x) = \frac{2}{\sqrt{\pi}} \int_0^x dt\, t^{1/2} e^{-t}; \qquad \psi'(x) = \frac{d\psi}{dx},$$

and $\lambda_{\alpha\beta} = \ln \Lambda_{\alpha\beta}$ is the Coulomb logarithm (see below). Limiting forms of ν_s, $\nu_\perp$ and $\nu_\parallel$ are given in the following table. All the expressions shown have units $cm^3 \, sec^{-1}$. Test particle energy ϵ and field particle temperature T are both in eV; $\mu = m_i/m_p$ where m_p is the proton mass; Z is ion charge state; in electron–electron and ion–ion encounters, field particle quantities are distinguished by a prime. The two expressions given below for each rate hold for very slow ($x^{\alpha\backslash\beta} \ll 1$) and very fast ($x^{\alpha\backslash\beta} \gg 1$) test particles, respectively.

	Slow	Fast

Electron–electron

$$\nu_s^{e|e}/n_e\lambda_{ee} \approx 5.8 \times 10^{-6}T^{-3/2} \longrightarrow 7.7 \times 10^{-6}\epsilon^{-3/2}$$

$$\nu_\perp^{e|e}/n_e\lambda_{ee} \approx 5.8 \times 10^{-6}T^{-1/2}\epsilon^{-1} \longrightarrow 7.7 \times 10^{-6}\epsilon^{-3/2}$$

$$\nu_\parallel^{e|e}/n_e\lambda_{ee} \approx 2.9 \times 10^{-6}T^{-1/2}\epsilon^{-1} \longrightarrow 3.9 \times 10^{-6}T\epsilon^{-5/2}$$

Electron–ion

$$\nu_s^{e|i}/n_iZ^2\lambda_{ei} \approx 0.23\mu^{3/2}T^{-3/2} \longrightarrow 3.9 \times 10^{-6}\epsilon^{-3/2}$$

$$\nu_\perp^{e|i}/n_iZ^2\lambda_{ei} \approx 2.5 \times 10^{-4}\mu^{1/2}T^{-1/2}\epsilon^{-1} \longrightarrow 7.7 \times 10^{-6}\epsilon^{-3/2}$$

$$\nu_\parallel^{e|i}/n_iZ^2\lambda_{ei} \approx 1.2 \times 10^{-4}\mu^{1/2}T^{-1/2}\epsilon^{-1} \longrightarrow 2.1 \times 10^{-9}\mu^{-1}T\epsilon^{-5/2}$$

Ion–electron

$$\nu_s^{i|e}/n_eZ^2\lambda_{ie} \approx 1.6 \times 10^{-9}\mu^{-1}T^{-3/2} \longrightarrow 1.7 \times 10^{-4}\mu^{1/2}\epsilon^{-3/2}$$

$$\nu_\perp^{i|e}/n_eZ^2\lambda_{ie} \approx 3.2 \times 10^{-9}\mu^{-1}T^{-1/2}\epsilon^{-1} \longrightarrow 1.8 \times 10^{-7}\mu^{-1/2}\epsilon^{-3/2}$$

$$\nu_\parallel^{i|e}/n_eZ^2\lambda_{ie} \approx 1.6 \times 10^{-9}\mu^{-1}T^{-1/2}\epsilon^{-1} \longrightarrow 1.7 \times 10^{-4}\mu^{1/2}T\epsilon^{-5/2}$$

Ion–ion

$$\frac{\nu_s^{i|i'}}{n_{i'}Z^2Z'^2\lambda_{ii'}} \approx 6.8 \times 10^{-8}\frac{\mu'^{1/2}}{\mu}\left(1+\frac{\mu'}{\mu}\right)^{-1/2}T^{-3/2}$$

$$\longrightarrow 9.0 \times 10^{-8}\left(\frac{1}{\mu}+\frac{1}{\mu'}\right)\frac{\mu^{1/2}}{\epsilon^{3/2}}$$

$$\frac{\nu_\perp^{i|i'}}{n_{i'}Z^2Z'^2\lambda_{ii'}} \approx 1.4 \times 10^{-7}\mu'^{1/2}\mu^{-1}T^{-1/2}\epsilon^{-1}$$

$$\longrightarrow 1.8 \times 10^{-7}\mu^{-1/2}\epsilon^{-3/2}$$

$$\frac{\nu_\parallel^{i|i'}}{n_{i'}Z^2Z'^2\lambda_{ii'}} \approx 6.8 \times 10^{-8}\mu'^{1/2}\mu^{-1}T^{-1/2}\epsilon^{-1}$$

$$\longrightarrow 9.0 \times 10^{-8}\mu^{1/2}\mu'^{-1}T\epsilon^{-5/2}$$

In the same limits, the energy transfer rate follows from the identity

$$\nu_\epsilon = 2\nu_s - \nu_\perp - \nu_\parallel,$$

except for the case of fast electrons or fast ions scattered by ions, where the leading terms cancel. Then the appropriate forms are

$$\nu_\epsilon^{e|i} \longrightarrow 4.2 \times 10^{-9}n_iZ^2\lambda_{ei}$$
$$\left[\epsilon^{-3/2}\mu^{-1} - 8.9 \times 10^4(\mu/T)^{1/2}\epsilon^{-1}\exp(-1836\mu\epsilon/T)\right]\,sec^{-1}$$

and

$$\nu_\epsilon^{i|i'} \longrightarrow 1.8 \times 10^{-7} n_{i'} Z^2 Z'^2 \lambda_{ii'}$$

$$\left[\epsilon^{-3/2}\mu^{1/2}/\mu' - 1.1[(\mu+\mu')/\mu\mu'](\mu'/T')^{1/2}\epsilon^{-1}\exp(-\mu'\epsilon/\mu T')\right] \text{ sec}^{-1}.$$

In general, the energy transfer rate $\nu_\epsilon^{\alpha\backslash\beta}$ is positive for $\epsilon > \epsilon_\alpha^*$ and negative for $\epsilon < \epsilon_\alpha^*$, where $x^* = (m_\beta/m_\alpha)\epsilon_\alpha^*/T_\beta$ is the solution of $\psi'(x^*) = (m_\alpha|m_\beta)\psi(x^*)$. The ratio $\epsilon_\alpha^*/T_\beta$ is given for a number of specific α, β in the following table:

$\alpha\backslash\beta$	$i\|e$	$e\|e, i\|i$	$e\|p$	$e\|D$	$e\|T, e\|He^3$	$e\|He^4$
$\dfrac{\epsilon_\alpha^*}{T_\beta}$	1.5	0.98	4.8×10^{-3}	2.6×10^{-3}	1.8×10^{-3}	1.4×10^{-3}

When both species are near Maxwellian, with $T_i \lesssim T_e$, there are just two characteristic collision rates. For $Z = 1$,

$$\nu_e = 2.9 \times 10^{-6} n\lambda T_e^{-3/2} \text{ sec}^{-1};$$

$$\nu_i = 4.8 \times 10^{-8} n\lambda T_i^{-3/2}\mu^{-1/2} \text{ sec}^{-1}.$$

Temperature Isotropization

Isotropization is described by

$$\frac{dT_\perp}{dt} = -\frac{1}{2}\frac{dT_\|}{dt} = -\nu_T^\alpha(T_\perp - T_\|),$$

where, if $A \equiv T_\perp/T_\| - 1 > 0$,

$$\nu_T^\alpha = \frac{2\sqrt{\pi}e_\alpha^2 e_\beta^2 n_\alpha \lambda_{\alpha\beta}}{m_\alpha^{1/2}(kT_\|)^{3/2}} A^{-2}\left[-3 + (A+3)\frac{\tan^{-1}(A^{1/2})}{A^{1/2}}\right].$$

If $A < 0$, $\tan^{-1}(A^{1/2})/A^{1/2}$ is replaced by $\tanh^{-1}(-A)^{1/2}/(-A)^{1/2}$. For $T_\perp \approx T_\| \equiv T$,

$$\nu_T^e = 8.2 \times 10^{-7} n\lambda T^{-3/2} \text{ sec}^{-1};$$

$$\nu_T^i = 1.9 \times 10^{-8} n\lambda Z^2 \mu^{-1/2} T^{-3/2} \text{ sec}^{-1}.$$

Thermal Equilibration

If the components of a plasma have different temperatures, but no relative drift, equilibration is described by

$$\frac{dT_\alpha}{dt} = \sum_\beta \bar{\nu}_\epsilon^{\alpha \backslash \beta} (T_\beta - T_\alpha),$$

where

$$\bar{\nu}_\epsilon^{\alpha \backslash \beta} = 1.8 \times 10^{-19} \frac{(m_\alpha m_\beta)^{1/2} Z_\alpha^2 Z_\beta^2 n_\beta \lambda_{\alpha\beta}}{(m_\alpha T_\beta + m_\beta T_\alpha)^{3/2}} \text{ sec}^{-1}.$$

For electrons and ions with $T_e \approx T_i \equiv T$, this implies

$$\bar{\nu}_\epsilon^{e|i}/n_i = \bar{\nu}_\epsilon^{i|e}/n_e = 3.2 \times 10^{-9} Z^2 \lambda / \mu T^{3/2} \text{cm}^3 \text{ sec}^{-1}.$$

Coulomb Logarithm

For test particles of mass m_α and charge $e_\alpha = Z_\alpha e$ scattering off field particles of mass m_β and charge $e_\beta = Z_\beta e$, the Coulomb logarithm is defined as $\lambda = \ln \Lambda \equiv \ln(r_{\max}/r_{\min})$. Here $r_{\min}$ is the larger of $e_\alpha e_\beta/m_{\alpha\beta}\bar{u}^2$ and $\hbar/2m_{\alpha\beta}\bar{u}$, averaged over both particle velocity distributions, where $m_{\alpha\beta} = m_\alpha m_\beta/(m_\alpha + m_\beta)$ and $\mathbf{u} = \mathbf{v}_\alpha - \mathbf{v}_\beta$; $r_{\max} = (4\pi \sum n_\gamma e_\gamma^2/kT_\gamma)^{-1/2}$, where the summation extends over all species γ for which $\bar{u}^2 < v_{T\gamma}^2 = kT_\gamma/m_\gamma$. If this inequality cannot be satisfied, or if either $\bar{u}\omega_{c\alpha}^{-1} < r_{\max}$ or $\bar{u}\omega_{c\beta}^{-1} < r_{\max}$, the theory breaks down. Typically $\lambda \approx 10\text{--}20$. Corrections to the transport coefficients are $O(\lambda^{-1})$; hence the theory is good only to $\sim 10\%$ and fails when $\lambda \sim 1$.

The following cases are of particular interest:

(a) Thermal electron–electron collisions

$$\lambda_{ee} = 23.5 - \ln(n_e^{1/2} T_e^{-5/4}) - [10^{-5} + (\ln T_e - 2)^2/16]^{1/2}$$

(b) Electron–ion collisions

$$\lambda_{ei} = \lambda_{ie} = 23 - \ln\left(n_e^{1/2} Z T_e^{-3/2}\right), \qquad T_i m_e/m_i < T_e < 10Z^2 \text{ eV};$$

$$= 24 - \ln\left(n_e^{1/2} T_e^{-1}\right), \qquad T_i m_e/m_i < 10Z^2 \text{ eV} < T_e$$

$$= 30 - \ln\left(n_i^{1/2} T_i^{-3/2} Z^2 \mu^{-1}\right), \qquad T_e < T_i Z m_e/m_i.$$

(c) Mixed ion–ion collisions

$$\lambda_{ii'} = \lambda_{i'i} = 23 - \ln\left[\frac{ZZ'(\mu + \mu')}{\mu T_{i'} + \mu' T_i}\left(\frac{n_i Z^2}{T_i} + \frac{n_{i'} Z'^2}{T_{i'}}\right)^{1/2}\right].$$

34

(d) Counterstreaming ions (relative velocity $v_D = \beta_D c$) in the presence of warm electrons, $kT_i/m_i, kT_{i'}/m_{i'} < v_D^2 < kT_e/m_e$

$$\lambda_{ii'} = \lambda_{i'i} = 35 - \ln\left[\frac{ZZ'(\mu + \mu')}{\mu\mu'\beta_D^2}\left(\frac{n_e}{T_e}\right)^{1/2}\right].$$

Fokker-Planck Equation

$$\frac{Df^\alpha}{Dt} \equiv \frac{\partial f^\alpha}{\partial t} + \mathbf{v} \cdot \nabla f^\alpha + \mathbf{F} \cdot \nabla_\mathbf{v} f^\alpha = \left(\frac{\partial f^\alpha}{\partial t}\right)_{\text{coll}},$$

where $\mathbf{F}$ is an external force field. The general form of the collision integral is $(\partial f^\alpha/\partial t)_{\text{coll}} = -\sum_\beta \nabla_\mathbf{v} \cdot \mathbf{J}^{\alpha\backslash\beta}$, with

$$\mathbf{J}^{\alpha\backslash\beta} = 2\pi\lambda_{\alpha\beta}\frac{e_\alpha^2 e_\beta^2}{m_\alpha}\int d^3v'(u^2 I - \mathbf{uu})u^{-3}$$

$$\cdot \left\{\frac{1}{m_\beta}f^\alpha(\mathbf{v})\nabla_{\mathbf{v'}}f^\beta(\mathbf{v'}) - \frac{1}{m_\alpha}f^\beta(\mathbf{v'})\nabla_\mathbf{v}f^\alpha(\mathbf{v})\right\}$$

(Landau form) where $\mathbf{u} = \mathbf{v'} - \mathbf{v}$ and I is the unit dyad, or alternatively,

$$\mathbf{J}^{\alpha\backslash\beta} = 4\pi\lambda_{\alpha\beta}\frac{e_\alpha^2 e_\beta^2}{m_\alpha^2}\left\{f^\alpha(\mathbf{v})\nabla_\mathbf{v}H(\mathbf{v}) - \frac{1}{2}\nabla_\mathbf{v} \cdot \left[f^\alpha(\mathbf{v})\nabla_\mathbf{v}\nabla_\mathbf{v}G(\mathbf{v})\right]\right\},$$

where the Rosenbluth potentials are

$$G(\mathbf{v}) = \int f^\beta(\mathbf{v'})u\,d^3v'$$

$$H(\mathbf{v}) = \left(1 + \frac{m_\alpha}{m_\beta}\right)\int f^\beta(\mathbf{v'})u^{-1}d^3v'.$$

If species α is a weak beam (number and energy density small compared with background) streaming through a Maxwellian plasma, then

$$\mathbf{J}^{\alpha\backslash\beta} = -\frac{m_\alpha}{m_\alpha + m_\beta}\nu_s^{\alpha\backslash\beta}\mathbf{v}f^\alpha - \frac{1}{2}\nu_\parallel^{\alpha\backslash\beta}\mathbf{vv} \cdot \nabla_\mathbf{v}f^\alpha$$

$$- \frac{1}{4}\nu_\perp^{\alpha\backslash\beta}\left(v^2 I - \mathbf{vv}\right) \cdot \nabla_\mathbf{v}f^\alpha.$$

B-G-K Collision Operator

For distribution functions with no large gradients in velocity space, the Fokker-Planck collision terms can be approximated according to

$$\frac{Df_e}{Dt} = \nu_{ee}(F_e - f_e) + \nu_{ei}(\bar{F}_e - f_e);$$

$$\frac{Df_i}{Dt} = \nu_{ie}(\bar{F}_i - f_i) + \nu_{ii}(F_i - f_i).$$

The respective slowing-down rates $\nu_s^{\alpha \backslash \beta}$ given in the Relaxation Rate section above can be used for $\nu_{\alpha\beta}$, assuming slow ions and fast electrons, with ϵ replaced by T_α. (For ν_{ee} and ν_{ii}, one can equally well use $\nu_\perp$, and the result is insensitive to whether the slow- or fast-test-particle limit is employed.) The Maxwellians F_α and $\bar{F}_\alpha$ are given by

$$F_\alpha = n_\alpha \left(\frac{m_\alpha}{2\pi k T_\alpha}\right)^{3/2} \exp\left\{-\left[\frac{m_\alpha(\mathbf{v} - \mathbf{v}_\alpha)^2}{2kT_\alpha}\right]\right\};$$

$$\bar{F}_\alpha = n_\alpha \left(\frac{m_\alpha}{2\pi k \bar{T}_\alpha}\right)^{3/2} \exp\left\{-\left[\frac{m_\alpha(\mathbf{v} - \bar{\mathbf{v}}_\alpha)^2}{2k\bar{T}_\alpha}\right]\right\},$$

where n_α, $\mathbf{v}_\alpha$ and T_α are the number density, mean drift velocity, and effective temperature obtained by taking moments of f_α. Some latitude in the definition of $\bar{T}_\alpha$ and $\bar{\mathbf{v}}_\alpha$ is possible;[20] one choice is $\bar{T}_e = T_i$, $\bar{T}_i = T_e$, $\bar{\mathbf{v}}_e = \mathbf{v}_i$, $\bar{\mathbf{v}}_i = \mathbf{v}_e$.

Transport Coefficients

Transport equations for a multispecies plasma:

$$\frac{d^\alpha n_\alpha}{dt} + n_\alpha \nabla \cdot \mathbf{v}_\alpha = 0;$$

$$m_\alpha n_\alpha \frac{d^\alpha \mathbf{v}_\alpha}{dt} = -\nabla p_\alpha - \nabla \cdot P_\alpha + Z_\alpha e n_\alpha \left[\mathbf{E} + \frac{1}{c}\mathbf{v}_\alpha \times \mathbf{B}\right] + \mathbf{R}_\alpha;$$

$$\frac{3}{2}n_\alpha \frac{d^\alpha kT_\alpha}{dt} + p_\alpha \nabla \cdot \mathbf{v}_\alpha = -\nabla \cdot \mathbf{q}_\alpha - P_\alpha : \nabla\mathbf{v}_\alpha + Q_\alpha.$$

Here $d^\alpha/dt \equiv \partial/\partial t + \mathbf{v}_\alpha \cdot \nabla$; $p_\alpha = n_\alpha kT_\alpha$, where k is Boltzmann's constant; $\mathbf{R}_\alpha = \sum_\beta \mathbf{R}_{\alpha\beta}$ and $Q_\alpha = \sum_\beta Q_{\alpha\beta}$, where $\mathbf{R}_{\alpha\beta}$ and $Q_{\alpha\beta}$ are respectively the momentum and energy gained by the αth species through collisions with the βth; P_α is the stress tensor; and $\mathbf{q}_\alpha$ is the heat flow.

36

The transport coefficients in a simple two-component plasma (electrons and singly charged ions) are tabulated below. Here $\parallel$ and $\perp$ refer to the direction of the magnetic field $\mathbf{B} = \mathbf{b}B$; $\mathbf{u} = \mathbf{v}_e - \mathbf{v}_i$ is the relative streaming velocity; $n_e = n_i \equiv n$; $\mathbf{j} = -ne\mathbf{u}$ is the current; $\omega_{ce} = 1.76 \times 10^7 B \sec^{-1}$ and $\omega_{ci} = (m_e/m_i)\omega_{ce}$ are the electron and ion gyrofrequencies, respectively; and the basic collisional times are taken to be

$$\tau_e = \frac{3\sqrt{m_e}(kT_e)^{3/2}}{4\sqrt{2\pi}\,n\lambda e^4} = 3.44 \times 10^5 \frac{T_e^{3/2}}{n\lambda} \sec,$$

where λ is the Coulomb logarithm, and

$$\tau_i = \frac{3\sqrt{m_i}(kT_i)^{3/2}}{4\sqrt{\pi}n\,\lambda e^4} = 2.09 \times 10^7 \frac{T_i^{3/2}}{n\lambda}\mu^{1/2} \sec.$$

In the limit of large fields ($\omega_{c\alpha}\tau_\alpha \gg 1$, $\alpha = i, e$) the transport processes may be summarized as follows:[21]

momentum transfer $\quad \mathbf{R}_{ei} = -\mathbf{R}_{ie} \equiv \mathbf{R} = \mathbf{R_u} + \mathbf{R}_T$;

frictional force $\quad \mathbf{R_u} = ne(\mathbf{j}_\parallel/\sigma_\parallel + \mathbf{j}_\perp/\sigma_\perp)$;

electrical conductivities $\quad \sigma_\parallel = 1.96\sigma_\perp$; $\quad \sigma_\perp = ne^2\tau_e/m_e$;

thermal force $\quad \mathbf{R}_T = -0.71n\nabla_\parallel(kT_e) - \dfrac{3n}{2\omega_{ce}\tau_e}\mathbf{b}\times\nabla_\perp(kT_e)$;

ion heating $\quad Q_i = \dfrac{3m_e}{m_i}\dfrac{nk}{\tau_e}(T_e - T_i)$;

electron heating $\quad Q_e = -Q_i - \mathbf{R}\cdot\mathbf{u}$;

ion heat flux $\quad \mathbf{q}_i = -\kappa_\parallel^i\nabla_\parallel(kT_i) - \kappa_\perp^i\nabla_\perp(kT_i) + \kappa_\wedge^i\mathbf{b}\times\nabla_\perp(kT_i)$;

ion thermal conductivities $\quad \kappa_\parallel^i = 3.9\dfrac{nkT_i\tau_i}{m_i}$; $\quad \kappa_\perp^i = \dfrac{2nkT_i}{m_i\omega_{ci}^2\tau_i}$; $\quad \kappa_\wedge^i = \dfrac{5nkT_i}{2m_i\omega_{ci}}$;

electron heat flux $\quad \mathbf{q}_e = \mathbf{q}_u^e + \mathbf{q}_T^e$;

frictional heat flux $\quad \mathbf{q}_u^e = 0.71nkT_e\mathbf{u}_\parallel + \dfrac{3nkT_e}{2\omega_{ce}\tau_e}\mathbf{b}\times\mathbf{u}_\perp$;

thermal gradient heat flux $\quad \mathbf{q}_T^e = -\kappa_\parallel^e\nabla_\parallel(kT_e) - \kappa_\perp^e\nabla_\perp(kT_e) - \kappa_\wedge^e\mathbf{b}\times\nabla_\perp(kT_e)$;

electron thermal conductivities $\quad \kappa_\parallel^e = 3.2\dfrac{nkT_e\tau_e}{m_e}$; $\quad \kappa_\perp^e = 4.7\dfrac{nkT_e}{m_e\omega_{ce}^2\tau_e}$; $\quad \kappa_\wedge^e = \dfrac{5nkT_e}{2m_e\omega_{ce}}$;

stress tensor (either species)

$$P_{xx} = -\frac{\eta_0}{2}(W_{xx} + W_{yy}) - \frac{\eta_1}{2}(W_{xx} - W_{yy}) - \eta_3 W_{xy};$$

$$P_{yy} = -\frac{\eta_0}{2}(W_{xx} + W_{yy}) + \frac{\eta_1}{2}(W_{xx} - W_{yy}) + \eta_3 W_{xy};$$

$$P_{xy} = P_{yx} = -\eta_1 W_{xy} + \frac{\eta_3}{2}(W_{xx} - W_{yy});$$

$$P_{xz} = P_{zx} = -\eta_2 W_{xz} - \eta_4 W_{yz};$$

$$P_{yz} = P_{zy} = -\eta_2 W_{yz} + \eta_4 W_{xz};$$

$$P_{zz} = -\eta_0 W_{zz}$$

(here the z axis is defined parallel to **B**);

ion viscosity

$$\eta_0^i = 0.96 nkT_i\tau_i; \quad \eta_1^i = \frac{3nkT_i}{10\omega_{ci}^2\tau_i}; \quad \eta_2^i = \frac{6nkT_i}{5\omega_{ci}^2\tau_i};$$

$$\eta_3^i = \frac{nkT_i}{2\omega_{ci}}; \quad \eta_4^i = \frac{nkT_i}{\omega_{ci}};$$

electron viscosity

$$\eta_0^e = 0.73 nkT_e\tau_e; \quad \eta_1^e = 0.51\frac{nkT_e}{\omega_{ce}^2\tau_e}; \quad \eta_2^e = 2.0\frac{nkT_e}{\omega_{ce}^2\tau_e};$$

$$\eta_3^e = -\frac{nkT_e}{2\omega_{ce}}; \quad \eta_4^e = -\frac{nkT_e}{\omega_{ce}}.$$

For both species the rate-of-strain tensor is defined as

$$W_{jk} = \frac{\partial v_j}{\partial x_k} + \frac{\partial v_k}{\partial x_j} - \frac{2}{3}\delta_{jk}\nabla \cdot \mathbf{v}.$$

When **B** = 0 the following simplifications occur:

$$\mathbf{R}_u = ne\mathbf{j}/\sigma_\parallel; \quad \mathbf{R}_T = -0.71n\nabla(kT_e); \quad \mathbf{q}_i = -\kappa_\parallel^i\nabla(kT_i);$$

$$\mathbf{q}_u^e = 0.71nkT_e\mathbf{u}; \quad \mathbf{q}_T^e = -\kappa_\parallel^e\nabla(kT_e); \quad P_{jk} = -\eta_0 W_{jk}.$$

For $\omega_{ce}\tau_e \gg 1 \gg \omega_{ci}\tau_i$, the electrons obey the high-field expressions and the ions obey the zero-field expressions.

Collisional transport theory is applicable when (1) macroscopic time rates of change satisfy $d/dt \ll 1/\tau$, where τ is the longest collisional time scale, and (in the absence of a magnetic field) (2) macroscopic length scales L satisfy $L \gg l$, where $l = \bar{v}\tau$ is the mean free path. In a strong field, $\omega_{ce}\tau \gg 1$, condition (2) is replaced by $L_\parallel \gg l$ and $L_\perp \gg \sqrt{lr_e}$ ($L_\perp \gg r_e$ in a uniform field), where $L_\parallel$ is a macroscopic scale parallel to the field **B** and $L_\perp$ is the smaller of $B/|\nabla_\perp B|$ and the transverse plasma dimension. In addition, the standard transport coefficients are valid only when (3) the Coulomb logarithm satisfies $\Lambda \gg 1$; (4) the electron gyroradius satisfies $r_e \gg \lambda_D$, or $8\pi n_e m_e c^2 \gg B^2$; (5) relative drifts $\mathbf{u} = \mathbf{v}_\alpha - \mathbf{v}_\beta$ between two species are small compared with the thermal velocities, i.e., $u^2 \ll kT_\alpha/m_\alpha$, kT_β/m_β; and (6) anomalous transport processes owing to microinstabilities are negligible.

Weakly Ionized Plasmas

Collision frequency for scattering of charged particles of species α by neutrals is

$$\nu_\alpha = n_0 \sigma_s^{\alpha|0} (kT_\alpha/m_\alpha)^{1/2},$$

where n_0 is the neutral density and $\sigma_s^{\alpha\backslash 0}$ is the cross section, typically $\sim 5 \times 10^{-15}$ cm^2 and weakly dependent on temperature.

When the system is small compared with a Debye length, $L \ll \lambda_D$, the charged particle diffusion coefficients are

$$D_\alpha = kT_\alpha/m_\alpha \nu_\alpha,$$

In the opposite limit, both species diffuse at the ambipolar rate

$$D_A = \frac{\mu_i D_e - \mu_e D_i}{\mu_i - \mu_e} = \frac{(T_i + T_e)D_i D_e}{T_i D_e + T_e D_i},$$

where $\mu_\alpha = e_\alpha/m_\alpha \nu_\alpha$ is the mobility. The conductivity σ_α satisfies $\sigma_\alpha = n_\alpha e_\alpha \mu_\alpha$.

In the presence of a magnetic field $\mathbf{B}$ the scalars μ and σ become tensors,

$$\mathbf{J}^\alpha = \boldsymbol{\sigma}^\alpha \cdot \mathbf{E} = \sigma_\parallel^\alpha \mathbf{E}_\parallel + \sigma_\perp^\alpha \mathbf{E}_\perp + \sigma_\wedge^\alpha \mathbf{E} \times \mathbf{b},$$

where $\mathbf{b} = \mathbf{B}/B$ and

$$\sigma_\parallel^\alpha = n_\alpha e_\alpha{}^2/m_\alpha \nu_\alpha;$$

$$\sigma_\perp^\alpha = \sigma_\parallel^\alpha \nu_\alpha{}^2/(\nu_\alpha{}^2 + \omega_{c\alpha}^2);$$

$$\sigma_\wedge^\alpha = \sigma_\parallel^\alpha \nu_\alpha \omega_{c\alpha}/(\nu_\alpha{}^2 + \omega_{c\alpha}^2).$$

Here $\sigma_\perp$ and $\sigma_\wedge$ are the Pedersen and Hall conductivities, respectively.

39

APPROXIMATE MAGNITUDES
IN SOME TYPICAL PLASMAS

Plasma Type	n cm^{-3}	T eV	ω_{pe} sec^{-1}	λ_D cm	$n\lambda_D{}^3$	ν_{ei} sec^{-1}
Interstellar gas	1	1	6×10^4	7×10^2	4×10^8	7×10^{-5}
Gaseous nebula	10^3	1	2×10^6	20	8×10^6	6×10^{-2}
Solar Corona	10^9	10^2	2×10^9	2×10^{-1}	8×10^6	60
Diffuse hot plasma	10^{12}	10^2	6×10^{10}	7×10^{-3}	4×10^5	40
Solar atmosphere, gas discharge	10^{14}	1	6×10^{11}	7×10^{-5}	40	2×10^9
Warm plasma	10^{14}	10	6×10^{11}	2×10^{-4}	8×10^2	10^7
Hot plasma	10^{14}	10^2	6×10^{11}	7×10^{-4}	4×10^4	4×10^6
Thermonuclear plasma	10^{15}	10^4	2×10^{12}	2×10^{-3}	8×10^6	5×10^4
Theta pinch	10^{16}	10^2	6×10^{12}	7×10^{-5}	4×10^3	3×10^8
Dense hot plasma	10^{18}	10^2	6×10^{13}	7×10^{-6}	4×10^2	2×10^{10}
Laser Plasma	10^{20}	10^2	6×10^{14}	7×10^{-7}	40	2×10^{12}

The diagram (facing) gives comparable information in graphical form.[22]

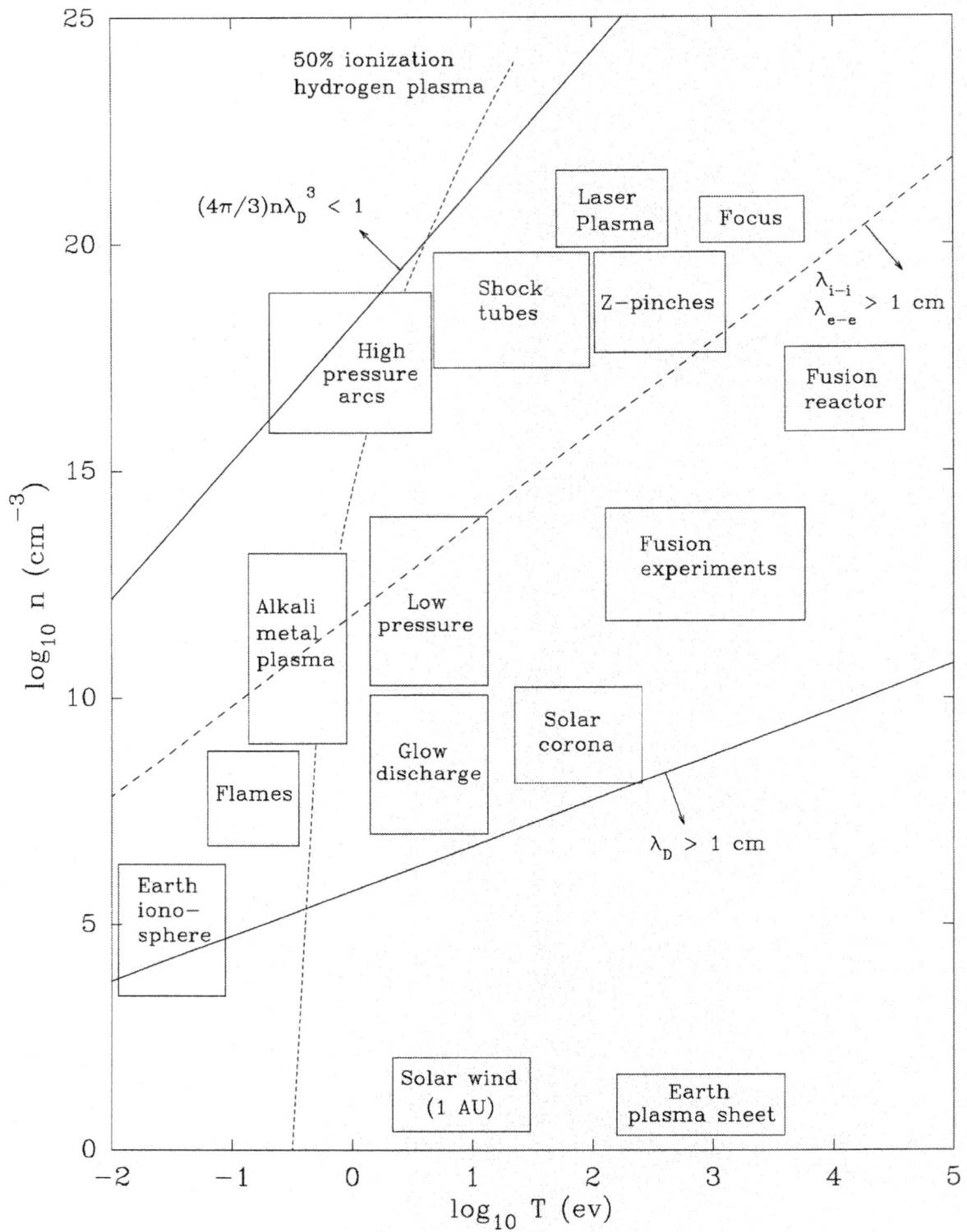

IONOSPHERIC PARAMETERS[23]

The following tables give average nighttime values. Where two numbers are entered, the first refers to the lower and the second to the upper portion of the layer.

Quantity	E Region	F Region
Altitude (km)	90–160	160–500
Number density (m^{-3})	1.5×10^{10}–3.0×10^{10}	5×10^{10}–2×10^{11}
Height-integrated number density (m^{-2})	9×10^{14}	4.5×10^{15}
Ion-neutral collision frequency (sec^{-1})	2×10^3–10^2	0.5–0.05
Ion gyro-/collision frequency ratio κ_i	0.09–2.0	4.6×10^2–5.0×10^3
Ion Pederson factor $\kappa_i/(1 + \kappa_i{}^2)$	0.09–0.5	2.2×10^{-3}–2×10^{-4}
Ion Hall factor $\kappa_i{}^2/(1 + \kappa_i{}^2)$	8×10^{-4}–0.8	1.0
Electron-neutral collision frequency	1.5×10^4–9.0×10^2	80–10
Electron gyro-/collision frequency ratio κ_e	4.1×10^2–6.9×10^3	7.8×10^4–6.2×10^5
Electron Pedersen factor $\kappa_e/(1 + \kappa_e{}^2)$	2.7×10^{-3}–1.5×10^{-4}	10^{-5}–1.5×10^{-6}
Electron Hall factor $\kappa_e{}^2/(1 + \kappa_e{}^2)$	1.0	1.0
Mean molecular weight	28–26	22–16
Ion gyrofrequency (sec^{-1})	180–190	230–300
Neutral diffusion coefficient ($m^2\ sec^{-1}$)	30–5×10^3	10^5

The terrestrial magnetic field in the lower ionosphere at equatorial lattitudes is approximately $B_0 = 0.35 \times 10^{-4}$ tesla. The earth's radius is $R_E = 6371$ km.

Parameter	Symbol	Value	Units
Total mass	$M_\odot$	1.99×10^{33}	g
Radius	$R_\odot$	6.96×10^{10}	cm
Surface gravity	$g_\odot$	2.74×10^4	$cm\,s^{-2}$
Escape speed	v_∞	6.18×10^7	$cm\,s^{-1}$
Upward mass flux in spicules	—	1.6×10^{-9}	$g\,cm^{-2}\,s^{-1}$
Vertically integrated atmospheric density	—	4.28	$g\,cm^{-2}$
Sunspot magnetic field strength	B_{max}	2500–3500	G
Surface effective temperature	T_0	5770	K
Radiant power	$\mathcal{L}_\odot$	3.83×10^{33}	$erg\,s^{-1}$
Radiant flux density	$\mathcal{F}$	6.28×10^{10}	$erg\,cm^{-2}s^{-1}$
Optical depth at 500 nm, measured from photosphere	τ_5	0.99	—
Astronomical unit (radius of earth's orbit)	AU	1.50×10^{13}	cm
Solar constant (intensity at 1 AU)	f	1.36×10^6	$erg\,cm^{-2}\,s^{-1}$

Chromosphere and Corona[25]

Parameter (Units)	Quiet Sun	Coronal Hole	Active Region
Chromospheric radiation losses ($erg\,cm^{-2}\,s^{-1}$)			
Low chromosphere	2×10^6	2×10^6	$\gtrsim 10^7$
Middle chromosphere	2×10^6	2×10^6	10^7
Upper chromosphere	3×10^5	3×10^5	2×10^6
Total	4×10^6	4×10^6	$\gtrsim 2 \times 10^7$
Transition layer pressure ($dyne\,cm^{-2}$)	0.2	0.07	2
Coronal temperature (K) at $1.1\,R_\odot$	$1.1\text{–}1.6 \times 10^6$	10^6	2.5×10^6
Coronal energy losses ($erg\,cm^{-2}\,s^{-1}$)			
Conduction	2×10^5	6×10^4	$10^5\text{–}10^7$
Radiation	10^5	10^4	5×10^6
Solar Wind	$\lesssim 5 \times 10^4$	7×10^5	$< 10^5$
Total	3×10^5	8×10^5	10^7
Solar wind mass loss ($g\,cm^{-2}\,s^{-1}$)	$\lesssim 2 \times 10^{-11}$	2×10^{-10}	$< 4 \times 10^{-11}$

THERMONUCLEAR FUSION[26]

Natural abundance of isotopes:

hydrogen	$n_D/n_H = 1.5 \times 10^{-4}$	
helium	$n_{He3}/n_{He4} = 1.3 \times 10^{-6}$	
lithium	$n_{Li6}/n_{Li7} = 0.08$	

Mass ratios:

$$m_e/m_D = 2.72 \times 10^{-4} = 1/3670$$
$$(m_e/m_D)^{1/2} = 1.65 \times 10^{-2} = 1/60.6$$
$$m_e/m_T = 1.82 \times 10^{-4} = 1/5496$$
$$(m_e/m_T)^{1/2} = 1.35 \times 10^{-2} = 1/74.1$$

Absorbed radiation dose is measured in rads: 1 rad $= 10^2 \, \mathrm{erg \, g^{-1}}$. The curie (abbreviated Ci) is a measure of radioactivity: 1 curie $= 3.7 \times 10^{10} \, \mathrm{counts \, sec^{-1}}$.

Fusion reactions (branching ratios are correct for energies near the cross section peaks; a negative yield means the reaction is endothermic):[27]

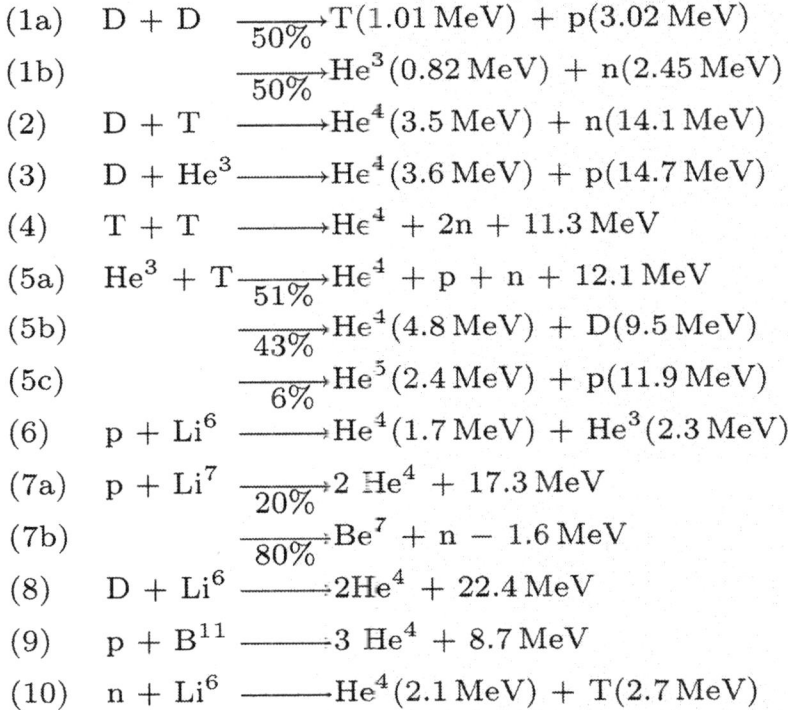

(1a) $D + D \xrightarrow[50\%]{} T(1.01\,\mathrm{MeV}) + p(3.02\,\mathrm{MeV})$

(1b) $\xrightarrow[50\%]{} He^3(0.82\,\mathrm{MeV}) + n(2.45\,\mathrm{MeV})$

(2) $D + T \longrightarrow He^4(3.5\,\mathrm{MeV}) + n(14.1\,\mathrm{MeV})$

(3) $D + He^3 \longrightarrow He^4(3.6\,\mathrm{MeV}) + p(14.7\,\mathrm{MeV})$

(4) $T + T \longrightarrow He^4 + 2n + 11.3\,\mathrm{MeV}$

(5a) $He^3 + T \xrightarrow[51\%]{} He^4 + p + n + 12.1\,\mathrm{MeV}$

(5b) $\xrightarrow[43\%]{} He^4(4.8\,\mathrm{MeV}) + D(9.5\,\mathrm{MeV})$

(5c) $\xrightarrow[6\%]{} He^5(2.4\,\mathrm{MeV}) + p(11.9\,\mathrm{MeV})$

(6) $p + Li^6 \longrightarrow He^4(1.7\,\mathrm{MeV}) + He^3(2.3\,\mathrm{MeV})$

(7a) $p + Li^7 \xrightarrow[20\%]{} 2\,He^4 + 17.3\,\mathrm{MeV}$

(7b) $\xrightarrow[80\%]{} Be^7 + n - 1.6\,\mathrm{MeV}$

(8) $D + Li^6 \longrightarrow 2He^4 + 22.4\,\mathrm{MeV}$

(9) $p + B^{11} \longrightarrow 3\,He^4 + 8.7\,\mathrm{MeV}$

(10) $n + Li^6 \longrightarrow He^4(2.1\,\mathrm{MeV}) + T(2.7\,\mathrm{MeV})$

The total cross section in barns (1 barn $= 10^{-24} \, \mathrm{cm}^2$) as a function of E, the energy in keV of the incident particle [the first ion on the left side of Eqs. (1)–(5)], assuming the target ion at rest, can be fitted by[28]

$$\sigma_T(E) = \frac{A_5 - \left[(A_4 - A_3 E)^2 + 1\right]^{-1} A_2}{E\left[\exp(A_1 E^{-1/2}) - 1\right]}$$

44

where the Duane coefficients A_j for the principle fusion reactions are as follows:

	D–D (1a)	D–D (1b)	D–T (2)	D–He^3 (3)	T–T (4)	T–He^3 (5a–c)
A_1	46.097	47.88	45.95	89.27	38.39	123.1
A_2	372	482	50200	25900	448	11250
A_3	4.36×10^{-4}	3.08×10^{-4}	1.368×10^{-2}	3.98×10^{-3}	1.02×10^{-3}	0
A_4	1.220	1.177	1.076	1.297	2.09	0
A_5	0	0	409	647	0	0

Reaction rates $\overline{\sigma v}$ (in cm^3 sec^{-1}), averaged over Maxwellian distributions:

Temperature (keV)	D–D (1a + 1b)	D–T (2)	D–He^3 (3)	T–T (4)	T–He^3 (5a–c)
1.0	1.5×10^{-22}	5.5×10^{-21}	10^{-26}	3.3×10^{-22}	10^{-28}
2.0	5.4×10^{-21}	2.6×10^{-19}	1.4×10^{-23}	7.1×10^{-21}	10^{-25}
5.0	1.8×10^{-19}	1.3×10^{-17}	6.7×10^{-21}	1.4×10^{-19}	2.1×10^{-22}
10.0	1.2×10^{-18}	1.1×10^{-16}	2.3×10^{-19}	7.2×10^{-19}	1.2×10^{-20}
20.0	5.2×10^{-18}	4.2×10^{-16}	3.8×10^{-18}	2.5×10^{-18}	2.6×10^{-19}
50.0	2.1×10^{-17}	8.7×10^{-16}	5.4×10^{-17}	8.7×10^{-18}	5.3×10^{-18}
100.0	4.5×10^{-17}	8.5×10^{-16}	1.6×10^{-16}	1.9×10^{-17}	2.7×10^{-17}
200.0	8.8×10^{-17}	6.3×10^{-16}	2.4×10^{-16}	4.2×10^{-17}	9.2×10^{-17}
500.0	1.8×10^{-16}	3.7×10^{-16}	2.3×10^{-16}	8.4×10^{-17}	2.9×10^{-16}
1000.0	2.2×10^{-16}	2.7×10^{-16}	1.8×10^{-16}	8.0×10^{-17}	5.2×10^{-16}

For low energies ($T \lesssim 25\,\text{keV}$) the data may be represented by

$$(\overline{\sigma v})_{DD} = 2.33 \times 10^{-14} T^{-2/3} \exp(-18.76 T^{-1/3})\ \text{cm}^3\,\text{sec}^{-1};$$

$$(\overline{\sigma v})_{DT} = 3.68 \times 10^{-12} T^{-2/3} \exp(-19.94 T^{-1/3})\ \text{cm}^3\,\text{sec}^{-1},$$

where T is measured in keV.

The power density released in the form of charged particles is

$$P_{DD} = 3.3 \times 10^{-13} n_D^2 (\overline{\sigma v})_{DD}\ \text{watt cm}^{-3} \quad \text{(including the subsequent D–T reaction)};$$

$$P_{DT} = 5.6 \times 10^{-13} n_D n_T (\overline{\sigma v})_{DT}\ \text{watt cm}^{-3};$$

$$P_{D\text{He}^3} = 2.9 \times 10^{-12} n_D\, n_{\text{He}^3} (\overline{\sigma v})_{D\text{He}^3}\ \text{watt cm}^{-3}.$$

45

RELATIVISTIC ELECTRON BEAMS

Here $\gamma = (1 - \beta^2)^{-1/2}$ is the relativistic scaling factor; quantities in analytic formulas are expressed in SI or cgs units, as indicated; in numerical formulas, I is in amperes (A), B is in gauss (G), electron linear density N is in cm^{-1}, and temperature, voltage and energy are in MeV; $\beta_z = v_z/c$; k is Boltzmann's constant.

Relativistic electron gyroradius:

$$r_e = \frac{mc^2}{eB}(\gamma^2 - 1)^{1/2} \text{ (cgs)} = 1.70 \times 10^3 (\gamma^2 - 1)^{1/2} B^{-1} \text{ cm.}$$

Relativistic electron energy:

$$W = mc^2\gamma = 0.511\gamma \text{ MeV.}$$

Bennett pinch condition:

$$I^2 = 2Nk(T_e + T_i)c^2 \text{ (cgs)} = 3.20 \times 10^{-4} N(T_e + T_i) \text{ A}^2.$$

Alfvén-Lawson limit:

$$I_A = (mc^3/e)\beta_z\gamma \text{ (cgs)} = (4\pi mc/\mu_0 e)\beta_z\gamma \text{ (SI)} = 1.70 \times 10^4 \beta_z\gamma \text{ A.}$$

The ratio of net current to I_A is

$$\frac{I}{I_A} = \frac{\nu}{\gamma}.$$

Here $\nu = Nr_e$ is the Budker number, where $r_e = e^2/mc^2 = 2.82 \times 10^{-13}$ cm is the classical electron radius. Beam electron number density is

$$n_b = 2.08 \times 10^8 J\beta^{-1} \text{ cm}^{-3},$$

where J is the current density in $A\,cm^{-2}$. For a uniform beam of radius a (in cm),

$$n_b = 6.63 \times 10^7 I a^{-2} \beta^{-1} \text{ cm}^{-3},$$

and

$$\frac{2r_e}{a} = \frac{\nu}{\gamma}.$$

Child's law: (non-relativistic) space-charge-limited current density between parallel plates with voltage drop V (in MV) and separation d (in cm) is

$$J = 2.34 \times 10^3 V^{3/2} d^{-2} \, \text{A cm}^{-2}.$$

The saturated parapotential current (magnetically self-limited flow along equipotentials in pinched diodes and transmission lines) is[29]

$$I_p = 8.5 \times 10^3 G\gamma \ln\left[\gamma + (\gamma^2 - 1)^{1/2}\right] \, \text{A},$$

where G is a geometrical factor depending on the diode structure:

$$G = \frac{w}{2\pi d} \qquad \text{for parallel plane cathode and anode of width } w, \text{ separation } d;$$

$$G = \left(\ln \frac{R_2}{R_1}\right)^{-1} \qquad \text{for cylinders of radii } R_1 \text{ (inner) and } R_2 \text{ (outer)};$$

$$G = \frac{R_c}{d_0} \qquad \text{for conical cathode of radius } R_c, \text{ maximum separation } d_0 \text{ (at } r = R_c\text{) from plane anode.}$$

For $\beta \to 0$ ($\gamma \to 1$), both I_A and I_p vanish.

The condition for a longitudinal magnetic field B_z to suppress filamentation in a beam of current density J (in A cm^{-2}) is

$$B_z > 47\beta_z(\gamma J)^{1/2} \, \text{G}.$$

Voltage registered by Rogowski coil of minor cross-sectional area A, n turns, major radius a, inductance L, external resistance R and capacitance C (all in SI):

externally integrated $V = (1/RC)(nA\mu_0 I/2\pi a)$;

self-integrating $V = (R/L)(nA\mu_0 I/2\pi a) = RI/n$.

X-ray production, target with average atomic number Z ($V \lesssim 5 \, \text{MeV}$):

$$\eta \equiv \text{x-ray power/beam power} = 7 \times 10^{-4} ZV.$$

X-ray dose at 1 meter generated by an e-beam depositing total charge Q coulombs while $V \geq 0.84 V_{\max}$ in material with charge state Z:

$$D = 150 V_{\max}^{2.8} Q Z^{1/2} \, \text{rads}.$$

47

BEAM INSTABILITIES[30]

Name	Conditions	Saturation Mechanism
Electron-electron	$V_d > \bar{V}_{ej},\ j = 1, 2$	Electron trapping until $\bar{V}_{ej} \sim V_d$
Buneman	$V_d > (M/m)^{1/3}\bar{V}_i,$ $V_d > \bar{V}_e$	Electron trapping until $\bar{V}_e \sim V_d$
Beam-plasma	$V_b > (n_p/n_b)^{1/3}\bar{V}_b$	Trapping of beam electrons
Weak beam-plasma	$V_b < (n_p/n_b)^{1/3}\bar{V}_b$	Quasilinear or nonlinear (mode coupling)
Beam-plasma (hot-electron)	$\bar{V}_e > V_b > \bar{V}_b$	Quasilinear or nonlinear
Ion acoustic	$T_e \gg T_i,\ V_d \gg C_s$	Quasilinear, ion tail formation, nonlinear scattering, or resonance broadening.
Anisotropic temperature (hydro)	$T_{e\perp} > 2T_{e\parallel}$	Isotropization
Ion cyclotron	$V_d > 20\bar{V}_i$ (for $T_e \approx T_i$)	Ion heating
Beam-cyclotron (hydro)	$V_d > C_s$	Resonance broadening
Modified two-stream (hydro)	$V_d < (1 + \beta)^{1/2}V_A,$ $V_d > C_s$	Trapping
Ion-ion (equal beams)	$U < 2(1 + \beta)^{1/2}V_A$	Ion trapping
Ion-ion (equal beams)	$U < 2C_s$	Ion trapping

For nomenclature, see p. 50.

Name	Parameters of Most Unstable Mode			
	Growth Rate	Frequency	Wave Number	Group Velocity
Electron-electron	$\frac{1}{2}\omega_e$	0	$0.9\dfrac{\omega_e}{V_d}$	0
Buneman	$0.7\left(\dfrac{m}{M}\right)^{1/3}\omega_e$	$0.4\left(\dfrac{m}{M}\right)^{1/3}\omega_e$	$\dfrac{\omega_e}{V_d}$	$\dfrac{2}{3}V_d$
Beam-plasma	$0.7\left(\dfrac{n_b}{n_p}\right)^{1/3}\omega_e$	$\omega_e - 0.4\left(\dfrac{n_b}{n_p}\right)^{1/3}\omega_e$	$\dfrac{\omega_e}{V_b}$	$\dfrac{2}{3}V_b$
Weak beam-plasma	$\dfrac{n_b}{2n_p}\left(\dfrac{V_b}{\bar V_b}\right)^2\omega_e$	ω_e	$\dfrac{\omega_e}{V_b}$	$\dfrac{3\bar V_e^2}{V_b}$
Beam-plasma (hot-electron)	$\left(\dfrac{n_b}{n_p}\right)^{1/2}\dfrac{\bar V_e}{V_b}\omega_e$	$\dfrac{V_b}{\bar V_e}\omega_e$	λ_D^{-1}	V_b
Ion acoustic	$\left(\dfrac{m}{M}\right)^{1/2}\omega_i$	ω_i	λ_D^{-1}	C_s
Anisotropic temperature (hydro)	Ω_e	$\omega_e\cos\theta\sim\Omega_e$	r_e^{-1}	$\bar V_{e\perp}$
Ion cyclotron	$0.1\Omega_i$	$1.2\Omega_i$	r_i^{-1}	$\dfrac{1}{3}\bar V_i$
Beam-cyclotron (hydro)	$0.7\Omega_e$	$n\Omega_e$	$0.7\lambda_D^{-1}$	$\gtrsim V_d;$ $\lesssim C_s$
Modified two-stream (hydro)	$\dfrac{1}{2}\Omega_H$	$0.9\Omega_H$	$1.7\dfrac{\Omega_H}{V_d}$	$\dfrac{1}{2}V_d$
Ion-ion (equal beams)	$0.4\Omega_H$	0	$1.2\dfrac{\Omega_H}{U}$	0
Ion-ion (equal beams)	$0.4\omega_i$	0	$1.2\dfrac{\omega_i}{U}$	0

For nomenclature, see p. 50.

In the preceding tables, subscripts e, i, d, b, p stand for "electron," "ion," "drift," "beam," and "plasma," respectively. Thermal velocities are denoted by a bar. In addition, the following are used:

m	electron mass	r_e, r_i	gyroradius
M	ion mass	β	plasma/magnetic energy
V	velocity		density ratio
T	temperature	V_A	Alfvén speed
n_e, n_i	number density	Ω_e, Ω_i	gyrofrequency
n	harmonic number	Ω_H	hybrid gyrofrequency,
$C_s = (T_e/M)^{1/2}$	ion sound speed		$\Omega_H{}^2 = \Omega_e \Omega_i$
ω_e, ω_i	plasma frequency	U	relative drift velocity of
λ_D	Debye length		two ion species

LASERS

System Parameters

Efficiencies and power levels are approximate.[31]

Type	Wavelength (μm)	Efficiency	Power levels available (W)	
			Pulsed	CW
CO_2	10.6	0.01–0.02 (pulsed)	$> 2 \times 10^{13}$	$> 10^5$
CO	5	0.4	$> 10^9$	> 100
Holmium	2.06	0.03†–0.1‡	$> 10^7$	80
Iodine	1.315	0.003	3×10^{12}	–
Nd-glass	1.06	–	1.25×10^{15}	–
Nd:YAG	1.064	–	10^9	$> 10^4$
Nd:YLF	1.045, 1.54, 1.313	–	4×10^8	80
Nd:YVO4	1.064	–	–	> 20
Er:YAG	2.94	–	1.5×10^5	–
*Color center	1–4	10^{-3}	5×10^8	1
*Ti:Sapphire	0.7–1.5	$0.4 \times \eta_p$	10^{14}	150
Ruby	0.6943	$< 10^{-3}$	10^{10}	1
He-Ne	0.6328	10^{-4}	–	1–50×10^{-3}
*Argon ion	0.45–0.60	10^{-3}	5×10^4	150
*OPO	0.3–10	$> 0.1 \times \eta_p$	10^{10}	5
N_2	0.3371	0.001–0.05	10^6	–
*Dye	0.3–1.1	10^{-3}	5×10^7	> 100
Kr-F	0.26	0.08	10^{12}	500
Xenon	0.175	0.02	$> 10^8$	–
Ytterbium fiber	1.05–1.1	0.55	5×10^7	10^4
Erbium fiber	1.534	–	7×10^6	100
Semiconductor	0.375–1.9	> 0.5	3×10^9	$> 10^3$

*Tunable sources   †lamp-driven   ‡diode-driven

Nd stands for Neodymium; Er stands for Erbium; Ti stands for Titanium; YAG stands for Yttrium–Aluminum Garnet; YLF stands for Yttrium Lithium Fluoride; YVO5 stands for Yttrium Vanadate; OPO for Optical Parametric Oscillator; η_p is pump laser efficiency.

51

Formulas

An e-m wave with $\mathbf{k} \parallel \mathbf{B}$ has an index of refraction given by

$$n_\pm = [1 - \omega_{pe}^2/\omega(\omega \mp \omega_{ce})]^{1/2},$$

where $\pm$ refers to the helicity. The rate of change of polarization angle θ as a function of displacement s (Faraday rotation) is given by

$$d\theta/ds = (k/2)(n_- - n_+) = 2.36 \times 10^4 N B f^{-2}\,\mathrm{cm}^{-1},$$

where N is the electron number density, B is the field strength, and f is the wave frequency, all in cgs.

The quiver velocity of an electron in an e-m field of angular frequency ω is

$$v_0 = eE_{\max}/m\omega = 25.6 I^{1/2}\lambda_0\,\mathrm{cm\,sec}^{-1}$$

in terms of the laser flux $I = cE_{\max}^2/8\pi$, with I in watt/cm^2, laser wavelength λ_0 in μm. The ratio of quiver energy to thermal energy is

$$W_{\mathrm{qu}}/W_{\mathrm{th}} = m_e v_0^2/2kT = 1.81 \times 10^{-13}\lambda_0^2 I/T,$$

where T is given in eV. For example, if $I = 10^{15}\,\mathrm{W\,cm}^{-2}$, $\lambda_0 = 1\,\mu$m, $T = 2\,$keV, then $W_{\mathrm{qu}}/W_{\mathrm{th}} \approx 0.1$.

Pondermotive force:

$$\mathcal{F} = N\nabla\langle E^2\rangle/8\pi N_c,$$

where

$$N_c = 1.1 \times 10^{21}\lambda_0^{-2}\mathrm{cm}^{-3}.$$

For uniform illumination of a lens with f-number F, the diameter d at focus (85% of the energy) and the depth of focus l (distance to first zero in intensity) are given by

$$d \approx 2.44F\lambda\theta/\theta_{DL} \quad \text{and} \quad l \approx \pm 2F^2\lambda\theta/\theta_{DL}.$$

Here θ is the beam divergence containing 85% of energy and θ_{DL} is the diffraction-limited divergence:

$$\theta_{DL} = 2.44\lambda/b,$$

where b is the aperture. These formulas are modified for nonuniform (such as Gaussian) illumination of the lens or for pathological laser profiles.

ATOMIC PHYSICS AND RADIATION

Energies and temperatures are in eV; all other units are cgs except where noted. Z is the charge state ($Z = 0$ refers to a neutral atom); the subscript e labels electrons. N refers to number density, n to principal quantum number. Asterisk superscripts on level population densities denote local thermodynamic equilibrium (LTE) values. Thus N_n^* is the LTE number density of atoms (or ions) in level n.

Characteristic atomic collision cross section:

$$(1) \qquad \pi a_0^2 = 8.80 \times 10^{-17} \, \text{cm}^2.$$

Binding energy of outer electron in level labelled by quantum numbers n, l:

$$(2) \qquad E_\infty^Z(n, l) = -\frac{Z^2 E_\infty^H}{(n - \Delta_l)^2},$$

where $E_\infty^H = 13.6 \, \text{eV}$ is the hydrogen ionization energy and $\Delta_l = 0.75 l^{-5}$, $l \gtrsim 5$, is the quantum defect.

Excitation and Decay

Cross section (Bethe approximation) for electron excitation by dipole allowed transition $m \rightarrow n$ (Refs. 32, 33):

$$(3) \qquad \sigma_{mn} = 2.36 \times 10^{-13} \frac{f_{mn} g(n, m)}{\epsilon \Delta E_{nm}} \, \text{cm}^2,$$

where f_{mn} is the oscillator strength, $g(n, m)$ is the Gaunt factor, ϵ is the incident electron energy, and $\Delta E_{nm} = E_n - E_m$.

Electron excitation rate averaged over Maxwellian velocity distribution, $X_{mn} = N_e \langle \sigma_{mn} v \rangle$ (Refs. 34, 35):

$$(4) \qquad X_{mn} = 1.6 \times 10^{-5} \frac{f_{mn} \langle g(n, m) \rangle N_e}{\Delta E_{nm} T_e^{1/2}} \exp\left(-\frac{\Delta E_{nm}}{T_e}\right) \, \text{sec}^{-1},$$

where $\langle g(n, m) \rangle$ denotes the thermal averaged Gaunt factor (generally ~ 1 for atoms, ~ 0.2 for ions).

Rate for electron collisional deexcitation:

$$(5) \qquad Y_{nm} = (N_m{}^*/N_n{}^*)X_{mn}.$$

Here $N_m{}^*/N_n{}^* = (g_m/g_n)\exp(\Delta E_{nm}/T_e)$ is the Boltzmann relation for level population densities, where g_n is the statistical weight of level n.

Rate for spontaneous decay $n \to m$ (Einstein A coefficient)[34]

$$(6) \qquad A_{nm} = 4.3 \times 10^7 (g_m/g_n)f_{mn}(\Delta E_{nm})^2 \ \sec^{-1}.$$

Intensity emitted per unit volume from the transition $n \to m$ in an optically thin plasma:

$$(7) \qquad I_{nm} = 1.6 \times 10^{-19} A_{nm} N_n \Delta E_{nm} \ \text{watt/cm}^3.$$

Condition for steady state in a corona model:

$$(8) \qquad N_0 N_e \langle \sigma_{0n} v \rangle = N_n A_{n0},$$

where the ground state is labelled by a zero subscript.

Hence for a transition $n \to m$ in ions, where $\langle g(n,0) \rangle \approx 0.2$,

$$(9) \qquad I_{nm} = 5.1 \times 10^{-25} \frac{f_{nm} g_m N_e N_0}{g_0 T_e^{1/2}} \left(\frac{\Delta E_{nm}}{\Delta E_{n0}} \right)^3 \exp \left(-\frac{\Delta E_{n0}}{T_e} \right) \frac{\text{watt}}{\text{cm}^3}.$$

Ionization and Recombination

In a general time-dependent situation the number density of the charge state Z satisfies

$$(10) \qquad \frac{dN(Z)}{dt} = N_e \Big[-S(Z)N(Z) - \alpha(Z)N(Z) $$
$$ +S(Z-1)N(Z-1) + \alpha(Z+1)N(Z+1) \Big].$$

Here $S(oZ)$ is the ionization rate. The recombination rate $\alpha(Z)$ has the form $\alpha(Z) = \alpha_r(Z) + N_e \alpha_3(Z)$, where α_r and α_3 are the radiative and three-body recombination rates, respectively.

Classical ionization cross-section[36] for any atomic shell j

$$\sigma_i = 6 \times 10^{-14} b_j g_j(x) / U_j^2 \text{ cm}^2.$$
(11)

Here b_j is the number of shell electrons; U_j is the binding energy of the ejected electron; $x = \epsilon / U_j$, where ϵ is the incident electron energy; and g is a universal function with a minimum value $g_{\min} \approx 0.2$ at $x \approx 4$.

Ionization from ion ground state, averaged over Maxwellian electron distribution, for $0.02 \lesssim T_e / E_\infty^Z \lesssim 100$ (Ref. 35):

$$S(Z) = 10^{-5} \frac{(T_e/E_\infty^Z)^{1/2}}{(E_\infty^Z)^{3/2}(6.0 + T_e/E_\infty^Z)} \exp\left(-\frac{E_\infty^Z}{T_e}\right) \text{ cm}^3/\text{sec},$$
(12)

where E_∞^Z is the ionization energy.

Electron-ion radiative recombination rate $(e + N(Z) \rightarrow N(Z-1) + h\nu)$ for $T_e/Z^2 \lesssim 400\,\text{eV}$ (Ref. 37):

$$\alpha_r(Z) = 5.2 \times 10^{-14} Z \left(\frac{E_\infty^Z}{T_e}\right)^{1/2} \left[0.43 + \frac{1}{2}\ln(E_\infty^Z/T_e)\right.$$
(13)

$$\left. + 0.469(E_\infty^Z/T_e)^{-1/3}\right] \text{ cm}^3/\text{sec}.$$

For $1\,\text{eV} < T_e/Z^2 < 15\,\text{eV}$, this becomes approximately[35]

$$\alpha_r(Z) = 2.7 \times 10^{-13} Z^2 T_e^{-1/2} \text{ cm}^3/\text{sec}.$$
(14)

Collisional (three-body) recombination rate for singly ionized plasma:[38]

$$\alpha_3 = 8.75 \times 10^{-27} T_e^{-4.5} \text{ cm}^6/\text{sec}.$$
(15)

Photoionization cross section for ions in level n, l (short-wavelength limit):

$$\sigma_{\text{ph}}(n, l) = 1.64 \times 10^{-16} Z^5 / n^3 K^{7+2l} \text{ cm}^2,$$
(16)

where K is the wavenumber in Rydbergs (1 Rydberg $= 1.0974 \times 10^5 \text{ cm}^{-1}$).

Ionization Equilibrium Models

Saha equilibrium:[39]

$$(17) \qquad \frac{N_e N_1{}^*(Z)}{N_n{}^*(Z-1)} = 6.0 \times 10^{21} \frac{g_1^Z T_e{}^{3/2}}{g_n^{Z-1}} \exp\left(-\frac{E_\infty^Z(n,l)}{T_e}\right) \text{cm}^{-3},$$

where g_n^Z is the statistical weight for level n of charge state Z and $E_\infty^Z(n,l)$ is the ionization energy of the neutral atom initially in level (n,l), given by Eq. (2).

In a steady state at high electron density,

$$(18) \qquad \frac{N_e N^*(Z)}{N^*(Z-1)} = \frac{S(Z-1)}{\alpha_3},$$

a function only of T.

Conditions for LTE:[39]

(a) Collisional and radiative excitation rates for a level n must satisfy

$$(19) \qquad Y_{nm} \gtrsim 10 A_{nm}.$$

(b) Electron density must satisfy

$$(20) \qquad N_e \gtrsim 7 \times 10^{18} Z^7 n^{-17/2} (T/E_\infty^Z)^{1/2} \text{cm}^{-3}.$$

Steady state condition in corona model:

$$(21) \qquad \frac{N(Z-1)}{N(Z)} = \frac{\alpha_r}{S(Z-1)}.$$

Corona model is applicable if[40]

$$(22) \qquad 10^{12} t_I{}^{-1} < N_e < 10^{16} T_e{}^{7/2} \text{cm}^{-3},$$

where t_I is the ionization time.

Radiation

N. B. Energies and temperatures are in eV; all other quantities are in cgs units except where noted. Z is the charge state ($Z = 0$ refers to a neutral atom); the subscript e labels electrons. N is number density.

Average radiative decay rate of a state with principal quantum number n is

$$(23) \qquad A_n = \sum_{m<n} A_{nm} = 1.6 \times 10^{10} Z^4 n^{-9/2} \text{ sec}.$$

Natural linewidth (ΔE in eV):

$$(24) \qquad \Delta E \, \Delta t = h = 4.14 \times 10^{-15} \text{ eV sec},$$

where Δt is the lifetime of the line.

Doppler width:

$$(25) \qquad \Delta\lambda/\lambda = 7.7 \times 10^{-5} (T/\mu)^{1/2},$$

where μ is the mass of the emitting atom or ion scaled by the proton mass.

Optical depth for a Doppler-broadened line:[39]

$$(26) \quad \tau = 3.52 \times 10^{-13} f_{nm} \lambda (Mc^2/kT)^{1/2} NL = 5.4 \times 10^{-9} f_{mn} \lambda (\mu/T)^{1/2} NL,$$

where f_{nm} is the absorption oscillator strength, λ is the wavelength, and L is the physical depth of the plasma; M, N, and T are the mass, number density, and temperature of the absorber; μ is M divided by the proton mass. Optically thin means $\tau < 1$.

Resonance absorption cross section at center of line:

$$(27) \qquad \sigma_{\lambda=\lambda_c} = 5.6 \times 10^{-13} \lambda^2/\Delta\lambda \text{ cm}^2.$$

Wien displacement law (wavelength of maximum black-body emission):

$$(28) \qquad \lambda_{\max} = 2.50 \times 10^{-5} T^{-1} \text{ cm}.$$

Radiation from the surface of a black body at temperature T:

$$(29) \qquad W = 1.03 \times 10^5 T^4 \text{ watt/cm}^2.$$

Bremsstrahlung from hydrogen-like plasma:[26]

$$(30) \qquad P_{\mathrm{Br}} = 1.69 \times 10^{-32} N_e T_e^{1/2} \sum \left[Z^2 N(Z) \right] \mathrm{watt/cm}^3,$$

where the sum is over all ionization states Z.

Bremsstrahlung optical depth:[41]

$$(31) \qquad \tau = 5.0 \times 10^{-38} N_e N_i Z^2 \overline{g} L T^{-7/2},$$

where $\overline{g} \approx 1.2$ is an average Gaunt factor and L is the physical path length.

Inverse bremsstrahlung absorption coefficient[42] for radiation of angular frequency ω:

$$(32) \qquad \kappa = 3.1 \times 10^{-7} Z n_e^2 \ln \Lambda \, T^{-3/2} \omega^{-2} (1 - \omega_p^2/\omega^2)^{-1/2} \mathrm{cm}^{-1};$$

here Λ is the electron thermal velocity divided by V, where V is the larger of ω and ω_p multiplied by the larger of Ze^2/kT and $\hbar/(mkT)^{1/2}$.

Recombination (free-bound) radiation:

$$(33) \qquad P_r = 1.69 \times 10^{-32} N_e T_e^{-1/2} \sum \left[Z^2 N(Z) \left(\frac{E_\infty^{Z-1}}{T_e} \right) \right] \mathrm{watt/cm}^3.$$

Cyclotron radiation[26] in magnetic field $\mathbf{B}$:

$$(34) \qquad P_c = 6.21 \times 10^{-28} B^2 N_e T_e \, \mathrm{watt/cm}^3.$$

For $N_e k T_e = N_i k T_i = B^2/16\pi$ ($\beta = 1$, isothermal plasma),[26]

$$(35) \qquad P_c = 5.00 \times 10^{-38} N_e^2 T_e^2 \, \mathrm{watt/cm}^3.$$

Cyclotron radiation energy loss e-folding time for a single electron:[41]

$$(36) \qquad t_c \approx \frac{9.0 \times 10^8 B^{-2}}{2.5 + \gamma} \, \mathrm{sec},$$

where γ is the kinetic plus rest energy divided by the rest energy mc^2.

Number of cyclotron harmonics[41] trapped in a medium of finite depth L:

$$(37) \qquad m_{\mathrm{tr}} = (57 \beta B L)^{1/6},$$

where $\beta = 8\pi N k T/B^2$.

Line radiation is given by summing Eq. (9) over all species in the plasma.

ATOMIC SPECTROSCOPY

Spectroscopic notation combines observational and theoretical elements. Observationally, spectral lines are grouped in series with line spacings which decrease toward the series limit. Every line can be related theoretically to a transition between two atomic states, each identified by its quantum numbers.

Ionization levels are indicated by roman numerals. Thus C I is unionized carbon, C II is singly ionized, etc. The state of a one-electron atom (hydrogen) or ion (He II, Li III, etc.) is specified by identifying the principal quantum number $n = 1, 2, \ldots$, the orbital angular momentum $l = 0, 1, \ldots, n-1$, and the spin angular momentum $s = \pm\frac{1}{2}$. The total angular momentum j is the magnitude of the vector sum of $\mathbf{l}$ and $\mathbf{s}$, $j = l \pm \frac{1}{2}$ ($j \geq \frac{1}{2}$). The letters s, p, d, f, g, h, i, k, l, $\ldots$, respectively, are associated with angular momenta $l = 0, 1, 2, 3, 4, 5, 6, 7, 8, \ldots$. The atomic states of hydrogen and hydrogenic ions are degenerate: neglecting fine structure, their energies depend only on n according to

$$E_n = -\frac{R_\infty h c Z^2 n^{-2}}{1 + m/M} = -\frac{\mathrm{Ry} Z^2}{n^2},$$

where h is Planck's constant, c is the velocity of light, m is the electron mass, M and Z are the mass and charge state of the nucleus, and

$$R_\infty = 109,737 \, \mathrm{cm}^{-1}$$

is the Rydberg constant. If E_n is divided by hc, the result is in wavenumber units. The energy associated with a transition $m \to n$ is given by

$$\Delta E_{mn} = \mathrm{Ry}(1/m^2 - 1/n^2),$$

with $m < n$ ($m > n$) for absorption (emission) lines.

For hydrogen and hydrogenic ions the series of lines belonging to the transitions $m \to n$ have conventional names:

Transition	$1 \to n$	$2 \to n$	$3 \to n$	$4 \to n$	$5 \to n$	$6 \to n$
Name	Lyman	Balmer	Paschen	Brackett	Pfund	Humphreys

Successive lines in any series are denoted α, β, γ, etc. Thus the transition $1 \to 3$ gives rise to the Lyman-β line. Relativistic effects, quantum electrodynamic effects (e.g., the Lamb shift), and interactions between the nuclear magnetic moment and the magnetic field due to the electron produce small shifts and splittings, $\lesssim 10^{-2}\,\mathrm{cm}^{-1}$; these last are called "hyperfine structure."

In many-electron atoms the electrons are grouped in closed and open shells, with spectroscopic properties determined mainly by the outer shell. Shell energies depend primarily on n; the shells corresponding to $n = 1, 2, 3, \ldots$ are called K, L, M, etc. A shell is made up of subshells of different angular momenta, each labeled according to the values of n, l, and the number of electrons it contains out of the maximum possible number, $2(2l + 1)$. For example, $2\mathrm{p}^5$ indicates that there are 5 electrons in the subshell corresponding to $l = 1$ (denoted by p) and $n = 2$.

In the lighter elements the electrons fill up subshells within each shell in the order s, p, d, etc., and no shell acquires electrons until the lower shells are full. In the heavier elements this rule does not always hold. But if a particular subshell is filled in a noble gas, then the same subshell is filled in the atoms of all elements that come later in the periodic table. The ground state configurations of the noble gases are as follows:

He	$1\mathrm{s}^2$
Ne	$1\mathrm{s}^2 2\mathrm{s}^2 2\mathrm{p}^6$
Ar	$1\mathrm{s}^2 2\mathrm{s}^2 2\mathrm{p}^6 3\mathrm{s}^2 3\mathrm{p}^6$
Kr	$1\mathrm{s}^2 2\mathrm{s}^2 2\mathrm{p}^6 3\mathrm{s}^2 3\mathrm{p}^6 3\mathrm{d}^{10} 4\mathrm{s}^2 4\mathrm{p}^6$
Xe	$1\mathrm{s}^2 2\mathrm{s}^2 2\mathrm{p}^6 3\mathrm{s}^2 3\mathrm{p}^6 3\mathrm{d}^{10} 4\mathrm{s}^2 4\mathrm{p}^6 4\mathrm{d}^{10} 5\mathrm{s}^2 5\mathrm{p}^6$
Rn	$1\mathrm{s}^2 2\mathrm{s}^2 2\mathrm{p}^6 3\mathrm{s}^2 3\mathrm{p}^6 3\mathrm{d}^{10} 4\mathrm{s}^2 4\mathrm{p}^6 4\mathrm{d}^{10} 4\mathrm{f}^{14} 5\mathrm{s}^2 5\mathrm{p}^6 5\mathrm{d}^{10} 6\mathrm{s}^2 6\mathrm{p}^6$

Alkali metals (Li, Na, K, etc.) resemble hydrogen; their transitions are described by giving n and l in the initial and final states for the single outer (valence) electron.

For general transitions in most atoms the atomic states are specified in terms of the parity $(-1)^{\Sigma l_i}$ and the magnitudes of the orbital angular momentum $\mathbf{L} = \Sigma \mathbf{l}_i$, the spin $\mathbf{S} = \Sigma \mathbf{s}_i$, and the total angular momentum $\mathbf{J} = \mathbf{L} + \mathbf{S}$, where all sums are carried out over the unfilled subshells (the filled ones sum to zero). If a magnetic field is present the projections M_L, M_S, and M of $\mathbf{L}$, $\mathbf{S}$, and $\mathbf{J}$ along the field are also needed. The quantum numbers satisfy $|M_L| \leq L \leq \nu l$, $|M_S| \leq S \leq \nu/2$, and $|M| \leq J \leq L + S$, where ν is the number of electrons in the unfilled subshell. Upper-case letters S, P, D, etc., stand for $L = 0, 1, 2$, etc., in analogy with the notation for a single electron. For example, the ground state of Cl is described by $3\mathrm{p}^5\,{}^2\mathrm{P}^{\circ}_{3/2}$. The first part indicates that there are 5 electrons in the subshell corresponding to $n = 3$ and $l = 1$. (The closed inner subshells $1\mathrm{s}^2 2\mathrm{s}^2 2\mathrm{p}^6 3\mathrm{s}^2$, identical with the configuration of Mg, are usually omitted.) The symbol 'P' indicates that the angular momenta of the outer electrons combine to give $L = 1$. The prefix '2' represents the value of the multiplicity $2S + 1$ (the number of states with nearly the same energy), which is equivalent to specifying $S = \frac{1}{2}$. The subscript 3/2 is the value of J. The superscript 'o' indicates that the state has odd parity; it would be omitted if the state were even.

The notation for excited states is similar. For example, helium has a state $1s2s\ ^3S_1$ which lies $19.72\,\mathrm{eV}$ ($159,856\,\mathrm{cm}^{-1}$) above the ground state $1s^2\ ^1S_0$. But the two "terms" do not "combine" (transitions between them do not occur) because this would violate, e.g., the quantum-mechanical selection rule that the parity must change from odd to even or from even to odd. For electric dipole transitions (the only ones possible in the long-wavelength limit), other selection rules are that the value of l of only one electron can change, and only by $\Delta l = \pm 1$; $\Delta S = 0$; $\Delta L = \pm 1$ or 0; and $\Delta J = \pm 1$ or 0 (but $L = 0$ does not combine with $L = 0$ and $J = 0$ does not combine with $J = 0$). Transitions are possible between the helium ground state (which has $S = 0$, $L = 0$, $J = 0$, and even parity) and, e.g., the state $1s2p\ ^1P_1^o$ (with $S = 0$, $L = 1$, $J = 1$, odd parity, excitation energy $21.22\,\mathrm{eV}$). These rules hold accurately only for light atoms in the absence of strong electric or magnetic fields. Transitions that obey the selection rules are called "allowed"; those that do not are called "forbidden."

The amount of information needed to adequately characterize a state increases with the number of electrons; this is reflected in the notation. Thus[43] O II has an allowed transition between the states $2p^2 3p'$ $^2F_{7/2}^o$ and $2p^2(^1D)3d'\ ^2F_{7/2}$ (and between the states obtained by changing J from $7/2$ to $5/2$ in either or both terms). Here both states have two electrons with $n = 2$ and $l = 1$; the closed subshells $1s^2 2s^2$ are not shown. The outer ($n = 3$) electron has $l = 1$ in the first state and $l = 2$ in the second. The prime indicates that if the outermost electron were removed by ionization, the resulting ion would not be in its lowest energy state. The expression (^1D) give the multiplicity and total angular momentum of the "parent" term, i.e., the subshell immediately below the valence subshell; this is understood to be the same in both states. (Grandparents, etc., sometimes have to be specified in heavier atoms and ions.) Another example[43] is the allowed transition from $2p^2(^3P)3p\ ^2P_{1/2}^o$ (or $^2P_{3/2}^o$) to $2p^2(^1D)3d'\ ^2S_{1/2}$, in which there is a "spin flip" (from antiparallel to parallel) in the $n = 2, l = 1$ subshell, as well as changes from one state to the other in the value of l for the valence electron and in L.

The description of fine structure, Stark and Zeeman effects, spectra of highly ionized or heavy atoms, etc., is more complicated. The most important difference between optical and X-ray spectra is that the latter involve energy changes of the inner electrons rather than the outer ones; often several electrons participate.

COMPLEX (DUSTY) PLASMAS

Complex (dusty) plasmas (CDPs) may be regarded as a new and unusual state of matter. CDPs contain charged microparticles (dust grains) in addition to electrons, ions, and neutral gas. Electrostatic coupling between the grains can vary over a wide range, so that the states of CDPs can change from weakly coupled (gaseous) to crystalline. CDPs can be investigated at the kinetic level (individual particles are easily visualized and relevant time scales are accessible). CDPs are of interest as a non-Hamiltonian system of interacting particles and as a means to study generic fundamental physics of self-organization, pattern formation, phase transitions, and scaling. Their discovery has therefore opened new ways of precision investigations in many-particle physics.

Typical experimental dust properties

grain size (radius) $a \simeq 0.3 - 30 \ \mu$m, mass $m_d \sim 3 \times 10^{-7} - 3 \times 10^{-13}$ g, number density (in terms of the interparticle distance) $n_d \sim \Delta^{-3} \sim 10^3 - 10^7$ cm^{-3}, temperature $T_d \sim 3 \times 10^{-2} - 10^2$ eV.

Typical discharge (bulk) plasmas

gas pressure $p \sim 10^{-2} - 1$ Torr, $T_i \simeq T_n \simeq 3 \times 10^{-2}$ eV, $v_{T_i} \simeq 7 \times 10^4$ cm/s (Ar), $T_e \sim 0.3 - 3$ eV, $n_i \simeq n_e \sim 10^8 - 10^{10}$ cm^{-3}, screening length $\lambda_D \simeq \lambda_{Di} \sim 20 - 200 \ \mu$m, $\omega_{pi} \simeq 2 \times 10^6 - 2 \times 10^7$ s^{-1} (Ar). B fields up to $B \sim 3$ T.

Dimensionless

Havnes parameter	$P = |Z| n_d / n_e$
normalized charge	$z = |Z| e^2 / k T_e a$
dust-dust scattering parameter	$\beta_d = Z^2 e^2 / k T_d \lambda_D$
dust-plasma scattering parameter	$\beta_{e,i} = |Z| e^2 / k T_{e,i} \lambda_D$
coupling parameter	$\Gamma = (Z^2 e^2 / k T_d \Delta) \exp(-\Delta/\lambda_D)$
lattice parameter	$\kappa = \Delta / \lambda_D$
particle parameter	$\alpha = a / \Delta$
lattice magnetization parameter	$\mu = \Delta / r_d$

Typical experimental values: $P \sim 10^{-4} - 10^2, z \simeq 2 - 4$ ($Z \sim 10^3 - 10^5$ electron charges), $\Gamma < 10^3, \kappa \sim 0.3 - 10, \alpha \sim 10^{-4} - 3 \times 10^{-2}, \mu < 1$

Frequencies

dust plasma frequency	$\omega_{pd} = (4\pi Z^2 e^2 n_d / m_d)^{1/2}$
	$\simeq (|Z| \frac{P}{1+P} m_i / m_d)^{1/2} \omega_{pi}$
charge fluctuation frequency	$\omega_{\text{ch}} \simeq \frac{1+z}{\sqrt{2\pi}} (a/\lambda_D) \omega_{pi}$

62

dust-gas friction rate	$\nu_{nd} \sim 10a^2 p/m_d v_{T_n}$
dust gyrofrequency	$\omega_{cd} = ZeB/m_d c$

Velocities

dust thermal velocity	$v_{T_d} = (kT_d/m_d)^{1/2} \equiv [\frac{T_d}{T_i}\frac{m_i}{m_d}]^{1/2} v_{T_i}$
dust acoustic wave velocity	$C_{\mathrm{DA}} = \omega_{pd}\lambda_D$
	$\simeq (|Z|\frac{P}{1+P}m_i/m_d)^{1/2} v_{T_i}$
dust Alfvén wave velocity	$v_{Ad} = B/(4\pi n_d m_d)^{1/2}$
dust-acoustic Mach number	V/C_{DA}
dust magnetic Mach number	V/v_{Ad}
dust lattice (acoustic) wave velocity	$C_{\mathrm{DL}}^{l,t} = \omega_{pd}\lambda_D F_{l,t}(\kappa)$

The range of the dust-lattice wavenumbers is $K\Delta < \pi$ The functions $F_{l,t}(\kappa)$ for longitudinal and transverse waves can be approximated[44,45] with accuracy $< 1\%$ in the range $\kappa \leq 5$:

$$F_l \simeq 2.70\kappa^{1/2}(1 - 0.096\kappa - 0.004\kappa^2), \qquad F_t \simeq 0.51\kappa(1 - 0.039\kappa^2),$$

Lengths

frictional dissipation length	$L_\nu = v_{T_d}/\nu_{nd}$
dust Coulomb radius	$R_{Ce,i} = |Z|e^2/kT_{e,i}$
dust gyroradius	$r_d = v_{T_d}/\omega_{cd}$

Grain Charging

The charge evolution equation is $d|Z|/dt = I_i - I_e$. From orbital motion limited (OML) theory[46] in the collisionless limit $l_{en(in)} \gg \lambda_D \gg a$:

$$I_e = \sqrt{8\pi}a^2 n_e v_{T_e} \exp(-z), \qquad I_i = \sqrt{8\pi}a^2 n_i v_{T_i}\left(1 + \frac{T_e}{T_i}z\right).$$

Grains are charged negatively. The grain charge can vary in response to spatial and temporal variations of the plasma. Charge fluctuations are always present, with frequency ω_{ch}. Other charging mechanisms are photoemission, secondary emission, thermionic emission, field emission, etc. Charged dust grains change the plasma composition, keeping quasineutrality. A measure of this is the Havnes parameter $P = |Z|n_d/n_e$. The balance of I_e and I_i yields

$$\exp(-z) = \left(\frac{m_i}{m_e}\frac{T_i}{T_e}\right)^{1/2}\left(1 + \frac{T_e}{T_i}z\right)[1 + P(z)]$$

63

When the relative charge density of dust is large, $P \gg 1$, the grain charge Z monotonically decreases.

Forces and momentum transfer

In addition to the usual electromagnetic forces, grains in complex plasmas are also subject to: gravity force $\mathbf{F_g} = m_d \mathbf{g}$; thermophoretic force

$$\mathbf{F}_{th} = -\frac{4\sqrt{2\pi}}{15}(a^2/v_{T_n})\kappa_n \nabla T_n$$

(where κ_n is the coefficient of gas thermal conductivity); forces associated with the momentum transfer from other species, $\mathbf{F}_\alpha = -m_d \nu_{\alpha d} \mathbf{V}_{\alpha d}$, i.e., neutral, ion, and electron drag. For collisions between charged particles, two limiting cases are distinguished by the magnitude of the scattering parameter β_α. When $\beta_\alpha \ll 1$ the result is independent of the sign of the potential. When $\beta_\alpha \gg 1$, the results for repulsive and attractive interaction potentials are different. For typical complex plasmas the hierarchy of scattering parameters is $\beta_e(\sim 0.01 - 0.3) \ll \beta_i(\sim 1 - 30) \ll \beta_d(\sim 10^3 - 3 \times 10^4)$. The generic expressions for different types of collisions are[47]

$$\nu_{\alpha d} = (4\sqrt{2\pi}/3)(m_\alpha/m_d)a^2 n_\alpha v_{T_\alpha}\Phi_{\alpha d}$$

Electron-dust collisions

$$\Phi_{ed} \simeq \frac{1}{2}z^2\Lambda_{ed} \qquad \beta_e \ll 1$$

Ion-dust collisions

$$\Phi_{id} = \begin{cases} \frac{1}{2}z^2(T_e/T_i)^2\Lambda_{id} & \beta_i < 5 \\ 2(\lambda_D/a)^2(\ln^2\beta_i + 2\ln\beta_i + 2), & \beta_i > 13 \end{cases}$$

Dust-dust collisons

$$\Phi_{dd} = \begin{cases} z_d^2\Lambda_{dd} & \beta_d \ll 1 \\ (\lambda_D/a)^2[\ln 4\beta_d - \ln\ln 4\beta_d], & \beta_d \gg 1 \end{cases}$$

where $z_d \equiv Z^2 e^2/akT_d$.

For $\nu_{dd} \sim \nu_{nd}$ the complex plasma is in a two-phase state, and for $\nu_{nd} \gg \nu_{dd}$ we have merely tracer particles (dust-neutral gas interaction dominates). The momentum transfer cross section is proportional to the Coulomb logarithm $\Lambda_{\alpha d}$ when the Coulomb scattering theory is applicable. It is determined by integration over the impact parameters, from ρ_{min} to ρ_{max}. ρ_{min} is due to finite grain size and is given by OML theory. $\rho_{max} = \lambda_D$ for repulsive interaction (applicable for $\beta_\alpha \ll 1$), and $\rho_{max} = \lambda_D(1+2\beta_\alpha)^{1/2}$ for attractive interaction (applicable up to $\beta_\alpha < 5$).

64

For repulsive interaction (electron-dust and dust-dust)

$$\Lambda_{\alpha d} = z_\alpha \int_0^\infty e^{-z_\alpha x} \ln[1 + 4(\lambda_D/a_\alpha)^2 x^2] dx - 2z_\alpha \int_1^\infty e^{-z_\alpha x} \ln(2x - 1) dx,$$

where $z_e = z$, $a_e = a$, and $a_d = 2a$.

For ion-dust (attraction)

$$\Lambda_{id} \simeq z \int_0^\infty e^{-zx} \ln \left[\frac{1 + 2(T_i/T_e)(\lambda_D/a)x}{1 + 2(T_i/T_e)x} \right] dx.$$

For $\nu_{dd} \gg \nu_{nd}$ the complex plasma behaves like a one phase system (dust-dust interaction dominates).

Phase Diagram of Complex Plasmas

The figure below represents different "phase states" of CDPs as functions of the electrostatic coupling parameter Γ and κ or α, respectively. The vertical dashed line at $\kappa = 1$ conditionally divides the system into Coulomb and Yukawa parts. With respect to the usual plasma phase, in the diagram below the complex plasmas are "located" mostly in the strong coupling regime (equivalent to the top left corner).

Regions I (V) represent Coulomb (Yukawa) crystals, the crystallization condition is[48] $\Gamma > 106(1 + \kappa + \kappa^2/2)^{-1}$. Regions II (VI) are for Coulomb (Yukawa) non-ideal plasmas – the characteristic range of dust-dust interaction (in terms of the momentum transfer) is larger than the intergrain distance (in terms of the Wigner-Seitz radius), $(\sigma/\pi)^{1/2} > (4\pi/3)^{-1/3}\Delta$, which implies that the interaction is essentially multiparticle.

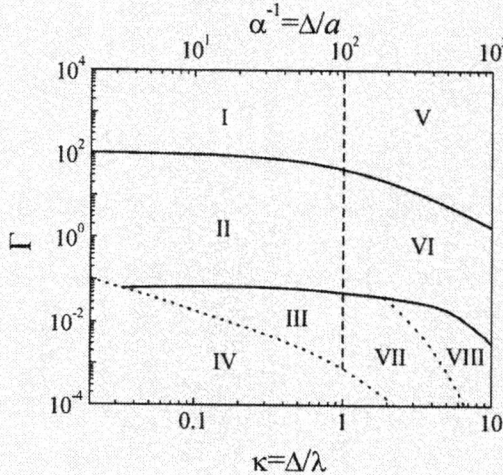

Regions III (VII and VIII) correspond to Coulomb (Yukawa) ideal gases. The range of dust-dust interaction is smaller than the intergrain distance and only pair collisions are important. In addition, in the region VIII the pair Yukawa interaction asymptotically reduces to the hard sphere limit, forming a "Yukawa granular medium". In region IV the electrostatic interaction is unimportant and the system is like a uaual granular medium.

65

REFERENCES

When any of the formulas and data in this collection are referenced in research publications, it is suggested that the original source be cited rather than the *Formulary*. Most of this material is well known and, for all practical purposes, is in the "public domain." Numerous colleagues and readers, too numerous to list by name, have helped in collecting and shaping the *Formulary* into its present form; they are sincerely thanked for their efforts.

Several book-length compilations of data relevant to plasma physics are available. The following are particularly useful:

C. W. Allen, *Astrophysical Quantities*, 3rd edition (Athlone Press, London, 1976).

A. Anders, *A Formulary for Plasma Physics* (Akademie-Verlag, Berlin, 1990).

H. L. Anderson (Ed.), *A Physicist's Desk Reference*, 2nd edition (American Institute of Physics, New York, 1989).

K. R. Lang, *Astrophysical Formulae*, 2nd edition (Springer, New York, 1980).

The books and articles cited below are intended primarily not for the purpose of giving credit to the original workers, but (1) to guide the reader to sources containing related material and (2) to indicate where to find derivations, explanations, examples, etc., which have been omitted from this compilation. Additional material can also be found in D. L. Book, NRL Memorandum Report No. 3332 (1977).

1. See M. Abramowitz and I. A. Stegun, Eds., *Handbook of Mathematical Functions* (Dover, New York, 1968), pp. 1–3, for a tabulation of some mathematical constants not available on pocket calculators.

2. H. W. Gould, "Note on Some Binomial Coefficient Identities of Rosenbaum," J. Math. Phys. **10**, 49 (1969); H. W. Gould and J. Kaucky, "Evaluation of a Class of Binomial Coefficient Summations," J. Comb. Theory **1**, 233 (1966).

3. B. S. Newberger, "New Sum Rule for Products of Bessel Functions with Application to Plasma Physics," J. Math. Phys. **23**, 1278 (1982); **24**, 2250 (1983).

4. P. M. Morse and H. Feshbach, *Methods of Theoretical Physics* (McGraw-Hill Book Co., New York, 1953), Vol. I, pp. 47–52 and pp. 656–666.

5. W. D. Hayes, "A Collection of Vector Formulas," Princeton University, Princeton, NJ, 1956 (unpublished), and personal communication (1977).

6. See *Quantities, Units and Symbols*, report of the Symbols Committee of the Royal Society, 2nd edition (Royal Society, London, 1975) for a discussion of nomenclature in SI units.

7. E. R. Cohen and B. N. Taylor, "The 1986 Adjustment of the Fundamental Physical Constants," CODATA Bulletin No. 63 (Pergamon Press, New York, 1986); J. Res. Natl. Bur. Stand. **92**, 85 (1987); J. Phys. Chem. Ref. Data **17**, 1795 (1988).

8. E. S. Weibel, "Dimensionally Correct Transformations between Different Systems of Units," Amer. J. Phys. **36**, 1130 (1968).

9. J. Stratton, *Electromagnetic Theory* (McGraw-Hill Book Co., New York, 1941), p. 508.

10. *Reference Data for Engineers: Radio, Electronics, Computer, and Communication*, 7th edition, E. C. Jordan, Ed. (Sams and Co., Indianapolis, IN, 1985), Chapt. 1. These definitions are International Telecommunications Union (ITU) Standards.

11. H. E. Thomas, *Handbook of Microwave Techniques and Equipment* (Prentice-Hall, Englewood Cliffs, NJ, 1972), p. 9. Further subdivisions are defined in Ref. 10, p. I–3.

12. J. P. Catchpole and G. Fulford, Ind. and Eng. Chem. **58**, 47 (1966); reprinted in recent editions of the *Handbook of Chemistry and Physics* (Chemical Rubber Co., Cleveland, OH) on pp. F306–323.

13. W. D. Hayes, "The Basic Theory of Gasdynamic Discontinuities," in *Fundamentals of Gas Dynamics*, Vol. III, *High Speed Aerodynamics and Jet Propulsion*, H. W. Emmons, Ed. (Princeton University Press, Princeton, NJ, 1958).

14. W. B. Thompson, *An Introduction to Plasma Physics* (Addison-Wesley Publishing Co., Reading, MA, 1962), pp. 86–95.

15. L. D. Landau and E. M. Lifshitz, *Fluid Mechanics*, 2nd edition (Addison-Wesley Publishing Co., Reading, MA, 1987), pp. 320–336.

16. The Z function is tabulated in B. D. Fried and S. D. Conte, *The Plasma Dispersion Function* (Academic Press, New York, 1961).

17. R. W. Landau and S. Cuperman, "Stability of Anisotropic Plasmas to Almost-Perpendicular Magnetosonic Waves," J. Plasma Phys. **6**, 495 (1971).

18. B. D. Fried, C. L. Hedrick, J McCune, "Two-Pole Approximation for the Plasma Dispersion Function," Phys. Fluids **11**, 249 (1968).

19. B. A. Trubnikov, "Particle Interactions in a Fully Ionized Plasma," *Reviews of Plasma Physics*, Vol. 1 (Consultants Bureau, New York, 1965), p. 105.

20. J. M. Greene, "Improved Bhatnagar–Gross–Krook Model of Electron-Ion Collisions," Phys. Fluids **16**, 2022 (1973).

21. S. I. Braginskii, "Transport Processes in a Plasma," *Reviews of Plasma Physics*, Vol. 1 (Consultants Bureau, New York, 1965), p. 205.

22. J. Sheffield, *Plasma Scattering of Electromagnetic Radiation* (Academic Press, New York, 1975), p. 6 (after J. W. Paul).

23. K. H. Lloyd and G. Härendel, "Numerical Modeling of the Drift and Deformation of Ionospheric Plasma Clouds and of their Interaction with Other Layers of the Ionosphere," J. Geophys. Res. **78**, 7389 (1973).

24. C. W. Allen, *Astrophysical Quantities*, 3rd edition (Athlone Press, London, 1976), Chapt. 9.

25. G. L. Withbroe and R. W. Noyes, "Mass and Energy Flow in the Solar Chromosphere and Corona," Ann. Rev. Astrophys. **15**, 363 (1977).

26. S. Glasstone and R. H. Lovberg, *Controlled Thermonuclear Reactions* (Van Nostrand, New York, 1960), Chapt. 2.

27. References to experimental measurements of branching ratios and cross sections are listed in F. K. McGowan, et al., Nucl. Data Tables **A6**, 353 (1969); **A8**, 199 (1970). The yields listed in the table are calculated directly from the mass defect.

28. G. H. Miley, H. Towner and N. Ivich, *Fusion Cross Section and Reactivities*, Rept. COO-2218-17 (University of Illinois, Urbana, IL, 1974); B. H. Duane, *Fusion Cross Section Theory*, Rept. BNWL-1685 (Brookhaven National Laboratory, 1972).

29. J. M. Creedon, "Relativistic Brillouin Flow in the High ν/γ Limit," J. Appl. Phys. **46**, 2946 (1975).

30. See, for example, A. B. Mikhailovskii, *Theory of Plasma Instabilities* Vol. I (Consultants Bureau, New York, 1974). The table on pp. 48–49 was compiled by K. Papadopoulos.

68

31. Table prepared from data compiled by J. M. McMahon (personal communication, D. Book, 1990) and A. Ting (personal communication, J.D. Huba, 2004).

32. M. J. Seaton, "The Theory of Excitation and Ionization by Electron Impact," in *Atomic and Molecular Processes*, D. R. Bates, Ed. (New York, Academic Press, 1962), Chapt. 11.

33. H. Van Regemorter, "Rate of Collisional Excitation in Stellar Atmospheres," Astrophys. J. **136**, 906 (1962).

34. A. C. Kolb and R. W. P. McWhirter, "Ionization Rates and Power Loss from θ-Pinches by Impurity Radiation," Phys. Fluids **7**, 519 (1964).

35. R. W. P. McWhirter, "Spectral Intensities," in *Plasma Diagnostic Techniques*, R. H. Huddlestone and S. L. Leonard, Eds. (Academic Press, New York, 1965).

36. M. Gryzinski, "Classical Theory of Atomic Collisions I. Theory of Inelastic Collision," Phys. Rev. **138A**, 336 (1965).

37. M. J. Seaton, "Radiative Recombination of Hydrogen Ions," Mon. Not. Roy. Astron. Soc. **119**, 81 (1959).

38. Ya. B. Zel'dovich and Yu. P. Raizer, *Physics of Shock Waves and High-Temperature Hydrodynamic Phenomena* (Academic Press, New York, 1966), Vol. I, p. 407.

39. H. R. Griem, *Plasma Spectroscopy* (Academic Press, New York, 1966).

40. T. F. Stratton, "X-Ray Spectroscopy," in *Plasma Diagnostic Techniques*, R. H. Huddlestone and S. L. Leonard, Eds. (Academic Press, New York, 1965).

41. G. Bekefi, *Radiation Processes in Plasmas* (Wiley, New York, 1966).

42. T. W. Johnston and J. M. Dawson, "Correct Values for High-Frequency Power Absorption by Inverse Bremsstrahlung in Plasmas," Phys. Fluids **16**, 722 (1973).

43. W. L. Wiese, M. W. Smith, and B. M. Glennon, *Atomic Transition Probabilities*, NSRDS-NBS 4, Vol. 1 (U.S. Govt. Printing Office, Washington, 1966).

44. F. M. Peeters and X. Wu, "Wigner crystal of a screened-Coulomb-interaction colloidal system in two dimensions", Phys. Rev. A **35**, 3109 (1987)

45. S. Zhdanov, R. A. Quinn, D. Samsonov, and G. E. Morfill, "Large-scale steady-state structure of a 2D plasma crystal", New J. Phys. **5**, 74 (2003).

46. J. E. Allen, "Probe theory – the orbital motion approach", Phys. Scripta **45**, 497 (1992).

47. S. A. Khrapak, A. V. Ivlev, and G. E. Morfill, "Momentum transfer in complex plasmas", Phys. Rev. E (2004).

48. V. E. Fortov *et al.*, "Dusty plasmas", Phys. Usp. **47**, 447 (2004).

AFTERWORD

The *NRL Plasma Formulary* originated over twenty five years ago and has been revised several times during this period. The guiding spirit and person primarily responsible for its existence is Dr. David Book. I am indebted to Dave for providing me with the TEX files for the *Formulary* and his continued suggestions for improvement. The *Formulary* has been set in TEX by Dave Book, Todd Brun, and Robert Scott. Finally, I thank readers for communicating typographical errors to me as well as suggestions for improvements.

www.ingramcontent.com/pod-product-compliance
Lightning Source LLC
Chambersburg PA
CBHW051421200326
41520CB00023B/7316